NOVEL AND RE-EMERGING RESPIRATORY VIRAL DISEASES

The Novartis Foundation is an international scientific and educational charity (UK Registered Charity No. 313574). Known until September 1997 as the Ciba Foundation, it was established in 1947 by the CIBA company of Basle, which merged with Sandoz in 1996, to form Novartis. The Foundation operates independently in London under English trust law. It was formally opened on 22 June 1949.

The Foundation promotes the study and general knowledge of science and in particular encourages international co-operation in scientific research. To this end, it organizes internationally acclaimed meetings (typically eight symposia and allied open meetings and 15–20 discussion meetings each year) and publishes eight books per year featuring the presented papers and discussions from the symposia. Although primarily an operational rather than a grant-making foundation, it awards bursaries to young scientists to attend the symposia and afterwards work with one of the other participants.

The Foundation's headquarters at 41 Portland Place, London W1B 1BN, provide library facilities, open to graduates in science and allied disciplines.

Towards the end of 2006, the Novartis Company undertook a review of the Foundation as a consequence of which the Foundation's Trustees were informed that Company support for the Foundation would cease with effect from the end of February 2008.

The Foundation's Trustees have considered various options for the future, the favoured of which is a merger with another, cognate, organization whereupon the Foundation will then formally be dissolved. Any future activities at 41 Portland Place will then be determined by the new organization.

Information on all Foundation activities can be found at http://www.novartisfound.org.uk

The Institute of Molecular and Cell Biology (IMCB) is a member of the Agency for Science, Technology and Research (A*STAR). Established in 1987, the research institute's mission is to foster a vibrant research culture for biomedical sciences and high quality manpower training to facilitate development of the biotechnology and pharmaceutical industries in Singapore.

Funded primarily by Biomedical Research Council (BMRC) of A*STAR, IMCB has about 35 core research labs and 8 core facility units consisting of over 400 research scientists. IMCB's research activities focus on five major fields: Cell Biology, Developmental Biology, Structural Biology, Infectious Diseases and Cancer Biology. IMCB continues to publish in renowned international journals, with more than a 1000 publications since 1987.

IMCB is currently based at The Biopolis @ One North. It is envisioned to be the biggest Biomedical Sciences R&D hub in Asia. IMCB continues to strive for excellence in biomedical R&D and the vision of Singapore as a world class hub for the Biomedical Sciences in Asia and beyond.

Novartis Foundation Symposium 290

NOVEL AND RE-EMERGING RESPIRATORY VIRAL DISEASES

John Wiley & Sons, Ltd

Published in 2008 by John Wiley & Sons Ltd,
The Atrium, Southern Gate,
Chichester PO19 8SQ, UK

National 01243 779777
International (+44) 1243 779777
e-mail (for orders and customer service enquiries); cs-books@wiley.co.uk
Visit our Home Page on http://www.wileyeurope.com
or http://www.wiley.com

Other Wiley Editorial Offices

John Wiley & Sons Inc., 111 River Street, Hoboken, NJ 07030, USA

Jossey-Bass, 989 Market Street, San Francisco, CA 94103-1741, USA

Wiley-VCH Verlag GmbH, Boschstr. 12, D-69469 Weinheim, Germany

John Wiley & Sons Australia Ltd, 33 Park Road, Milton, Queensland 4064, Australia

Jonh Wiley & Sons (Asia) Pte Ltd, 2 Clementi Loop #02-01, Jin Xing Distripark, Singapore 129809

John Wiley & Sons Canada Ltd, 6045 Freemont Blvd, Mississauga, Ontario, Canada L5R 4J3

Wiley also publishes its books in a variety of electronic formats. Some content that appears in print may not be available in electronic books.

Novartis Foundation Symposium 290

x + 156 pages, 9 figures, 2 plates, 12 tables

British Library Cataloguing in Publication Data

A catalogue record for this book is available from the British Library

ISBN 978-0-470-06538-9

Typeset in 10.5 on 12.5 pt Garamond by SNP Best-set Typesetter Ltd., Hong Kong
Printed and bound in Great Britain by T. J. International Ltd, Padstow, Cornwall.

Contents

Symposium on Novel and re-emerging respiratory viral diseases, held at the Institute of Molecular and Cell Biology, Singapore, 23–25 April 2007

Editors: Gregory Bock (Organizer) and Jamie Goode

This meeting was based on a proposal made by Yee-Joo Tan and Wanjin Hong

Participants

Larry J. Anderson Division of Viral Diseases, National Center for Immunizations and Respiratory Diseases, Centers for Disease Control and Prevention, 1600 Clifton Road, Atlanta, GA 30333, USA

Neal Greig Copeland Institute of Molecular and Cell Biology, 61 Biopolis Drive, Proteos, Singapore 138673

Martin Crusat The Oxford University Clinical Research Unit, Hospital for Tropical Diseases, 190 Ben Ham Tu, Quan 5, Ho Chi Minh City, Vietnam

Erich Hoffmann Department of Infectious Diseases, St. Jude Children's Research Hospital, 332 North Lauderdale, Memphis, TN 38105, USA

Edward C. Holmes Department of Biology, The Pennsylvania State University, Mueller Laboratory, University Park, PA 16802, USA

Wanjin Hong Institute of Molecular and Cell Biology, 61 Biopolis Drive, Proteos, Singapore 138673

Nancy Jenkins Institute of Molecular and Cell Biology, 61 Biopolis Drive, Proteos, Singapore 138673

Jeffrey S. Kahn Division of Infectious Diseases, Department of Pediatrics, Yale University School of Medicine, PO Box 208064, New Haven, CT 06520, USA

Yoshihiro Kawaoka Department of Pathobiological Sciences, 2015 Linden Drive, University of Wisconsin-Madison, Madison, WI 53706, USA and International Research Center for Infectious Diseases and Division of Virology, Department of Microbiology and Immunology, Institute of Medical Science, University of Tokyo, 4-6-1, Shirokanedai, Minato-ku, Tokyo 108-8639, Japan

Michael M. Lai Office of the Vice President, Academia Sinica, 128 Academia Road, Sec. 2, Nankang, Taipei 115, Taiwan

Sunil K. Lal Virology Group, International Centre for Genetic Engineering and Biotechnology, Aruna Asaf Ali Marg New Delhi 110067, India

David Lane Institute of Molecular and Cell Biology, 61 Biopolis Drive, Proteos, Singapore 138673

Yee Sin Leo Communicable Disease Centre, Tan Tock Seng Hospital, Moulmein Road, Singapore 308433

Ai Ee Ling Virology Laboratory, Department of Pathology, Singapore General Hospital, Outram Road, Singapore 168608

Ding Xiang Liu Institute of Molecular and Cell Biology, 61 Biopolis Drive, Proteos, Singapore 138673

Albert Osterhaus Department of Virology, Erasmus MC, Dr. Molewaterplein 50, PO Box 1738, 3000 DR Rotterdam, The Netherlands

J. S. Malik Peiris Department of Microbiology, The University of Hong Kong, Room 423, University Pathology Building, Queen Mary Hospital, Pokfulam, Hong Kong SAR

Shuo Shen Institute of Molecular and Cell Biology, 61 Biopolis Drive, Proteos, Singapore 138673

John J. Skehel MRC National Institute for Medical Research, The Ridgeway, Mill Hill, London NW7 1AA, UK

Derek J. Smith Department of Zoology, University of Cambridge, Downing Street, Cambridge CB2 3EJ, UK

Ih-Jen Su Division of Clinical Research, National Health Research Institutes, 138, Shen-Li Rd, Tainan, Taiwan

Paul Ananth Tambyah Division of Infectious Diseases, National University of Singapore, 5 Lower Kent Ridge Road, Singapore 119074

Yee-Joo Tan Institute of Molecular and Cell Biology, 61 Biopolis Drive, Proteos, Singapore 138673

Tran Tan Thanh The Oxford University Clinical Research Unit, Hospital for Tropical Diseases, 190 Ben Ham Tu, Quan 5, Ho Chi Minh City, Vietnam

Jean-Paul Thiery Institute of Molecular and Cell Biology, 61 Biopolis Drive, Proteos, Singapore 138673

Subhash Vasudevan Novartis Institute for Tropical Diseases, 10 Biopolis Road, Chromos #05-01, Singapore 138670

Robert G. Webster *(Chair)* Department of Infectious Diseases, Division of Virology, St. Jude Children's Research Hospital, 332 North Lauderdale, Memphis, TN 38105, USA

John M. Wood National Institute for Biological Standards and Control, Blanche Lane, South Mimms, Potters Bar, Herts EN6 3QG, UK

Li Xin Chinese National Influenza Center, National Institute for Viral Disease Control and Prevention, Chinese Center for Disease Control and Prevention, Beijing, China

Chair's introduction

Robert G. Webster

St. Jude Children's Research Hospital, Department of Infectious Diseases, Division of Virology, World Health Organization Collaborating Center for Studies on the Ecology of Influenza in Animals and Birds, 332 N. Lauderdale, Memphis, TN 38105, USA

Emerging and re-emerging infectious diseases are part of the natural history of humankind, for there has always been a struggle between microbes and humans. A considerable part of the human genome is concerned directly or indirectly with strategies to combat infectious diseases. Humans have continued their global dominance and in the past century have used scientific knowledge to reduce the impact of novel disease agents. The ever-increasing human population expansion and factors such as land use, water use and energy use needed to support the burgeoning human population, has resulted in production of animals on megafarms in close proximity to wild animals and birds. The export of intensive farming practices to the developing world, for example chicken and pig raising, has not always been accompanied by the best practices for ensuring bio-security and disease prevention in those operations. Thus intensive poultry and pig raising, without adequate separation from free-flying birds and water treatment, is a recipe for disaster. The increasing number of outbreaks of lethal H5 and H7 influenza, in domestic poultry, globally attests to these assertions.

The emergence of novel infectious diseases is a continuing process with multiple novel agents emerging in the past decade. While many of these agents caused transitory disease outbreaks—Nepah virus from bats to pigs and people in Malaysia, and Hendra virus from bats to horses and people in Australia—that were rapidly identified and stamped out, others became endemic in humans and in domestic animal species. Notable examples are human immunodeficiency virus (HIV) (African primates to humans) and West Nile virus (introduction to the Americas from Europe and spread through mosquitoes to wild birds, domestic mammals and humans).

Two recent examples of emerging infectious disease agents are severe acute respiratory syndrome (SARS) and highly pathogenic H5N1 avian influenza ('bird flu'). These two disease agents are the main topics for this meeting. Both of these diseases are caused by RNA viruses of zoonotic origin; SARS by a novel coronavirus from bats via civet cats in live animal markets ('wet markets') to humans, and H5N1 bird flu by a type A orthomyxovirus from wild aquatic birds via domestic poultry to humans. Both of these emerging infectious diseases were 'man made' in the sense that increased affluence of humans in the region increased the demand for protein in the diet. Intensified animal raising and the demand for exotic wild animal meat permitted these viruses to initially spread to humans in Hong Kong and Southern China through wet markets. The actual precursor viruses of neither SARS nor H5N1 bird influenza have been identified, but their closest genetic relatives were detected in animals and poultry in wet markets at the time they initially spread to humans.

Southeast Asia has been described as the epicentre for the emergence of pandemic influenza viruses, including the Asian H2N2 influenza of 1957, the Hong Kong H3N2 virus of 1968, as well as the re-emerging H1N1 Russian influenza virus of 1977. Both the H5N1 highly pathogenic avian influenza virus and the SARS coronavirus emerged in this region of the world. While culling of all domestic poultry in Hong Kong in 1997 successfully stamped out the initial genotype of H5N1, the virus re-emerged from apparently healthy ducks and geese in the region and spread to multiple countries in Southeast Asia including Vietnam, Cambodia, Laos, Indonesia, Japan and South Korea. The virus was largely restricted to the Southeast Asia region until 2005. The dramatic spread of the virus in mid-2005 occurred after H5N1 infected Bar-headed geese and other wild water fowl in Qinghai Lake in Western China. After that event, the virus spread rapidly through the Indian subcontinent, the African continent and Europe. The role of migratory birds seems probable. While the highly pathogenic H5N1 virus continues to spread throughout Eurasia it has, to date, not spread to the Americas despite the overlap of migrating birds in Alaska.

Both SARS and H5N1 bird flu are similar in being poorly transmissible in humans. During the SARS outbreak, this virus infected 8096 persons globally with 774 deaths (9.6%), while H5N1 bird flu has infected over 300 humans with 60% lethality. The poor transmissibility of SARS led to the control of this virus by conventional biosecurity and quarantine. While SARS is under control, H5N1 bird flu is not. H5N1 appeared in Hong Kong a decade ago: it has now spread to over 60 countries in Eurasia and has evolved into at least four antigenically distinct clades. Although H5N1 has not acquired consistent human-to-human transmission the possibility exists that we may be witnessing the evolution of a human influenza pandemic in real time.

Dr Yee-Joo Tan from The Institute of Molecular and Cell Biology, Proteos, Singapore, who participated in the battle against SARS in Singapore, proposed the topic of emerging and re-emerging respiratory viruses as the subject for the present meeting. Both the topic and the site for the meeting were most appropriate. Although the economic impact of SARS turned out to be relatively short term (due to rapid acquisition of scientific knowledge and control strategies) the initial impact on service exports in Singapore and Hong Kong, especially on tourism, was particularly severe. If SARS had not been controlled so expediently, the economic impact would have been much worse.

The lessons from SARS are certainly applicable to the expanding problem of H5N1 bird flu and to future emerging infectious diseases. The successful containment of SARS and the lessons learned from that successful programme are important to be considered in the face of a possibly emerging influenza pandemic in humans. However, we must keep in mind that the transmissibility of influenza is likely to be very different from that of the SARS coronavirus.

Identification and characterization of novel viruses

Larry J. Anderson and Suxiang Tong

Division of Viral Diseases, National Center for Immunizations and Respiratory Diseases, Centers for Disease Control and Prevention, Atlanta, GA 30333, USA[1]

Abstract. Although much has been learned about the agents and the clinical and epidemiological features of acute respiratory illness (ARI), much still remains unknown. Among children in the USA, the agent of 25–50% of cases of ARI remains unknown and among adults the agent remains unknown for about 50% of ARI cases. Roadblocks to detecting the etiological agent include quality of specimens, sensitivity and specificity of assays, and probably presence of as yet unknown pathogens. For example, since the year 2000, five new viral agents of ARI have been identified (human metapneunovirus [hMPV], SARS CoV, two human coronaviruses [NL63 and HKU1] and a new parvovirus [human bocavirus]). A variety of methods have been used to try to detect novel viral pathogens and include classic techniques such as tissue culture isolation, antigen detection assays and electron microscopy, and molecular methods designed specifically to detect novel pathogens. Examples of different successful methods to detect novel pathogens include those used to identify the hepatitis C virus, human herpes virus 8, Sin Nombre virus, and SARS coronavirus. At CDC, we have developed several molecular methods to identify new pathogens including pan viral family PCR assays that can detect any member of a given family of viruses. To date, we have developed pan viral family (or genera) PCR assays for 11 viral families. The improving methods to discover new viruses are likely to present investigators with the challenge to determine if and what disease an increasing number of novel viruses might cause. Koch's postulates provide guidelines for establishing a causal relationship between a pathogen and disease and include establishing an epidemiologic link between the pathogen and disease and then showing a causal relationship, most often through animal inoculation studies. The success of new molecular tools for pathogen discovery highlight the need for more efficient ways to determine what disease might be associated with the infection.

2008 Novel and re-emerging respiratory viral diseases. Wiley, Chichester (Novartis Foundation Symposium 290) p 4–16

Acute respiratory illnesses (ARI) include the common cold, bronchitis, bronchiolitis, croup, sore throat and pneumonia, and are the most common illnesses of humans. They are a major cause of morbidity and mortality world-wide. It is estimated that globally ~2 million deaths/year occur in children <5 years old from

[1]Disclaimer: The findings and conclusions in this report are those of the authors and do not necessarily represent the views of the Centers for Disease Control and Prevention/the Agency for Toxic Substances and Disease Registry.

TABLE 1 Pathogens associated with acute respiratory illness (ARI)

Viral agents
Adenoviruses
Coronaviruses including SARS CoV
Enteroviruses
Human bocavirus
Human metapneumovirus
Human parainfluenza viruses 1–4
Influenza virus A and B (Flu A and B)
Respiratory Syncytial Virus (RSV)
Rhinoviruses
Bacterial agents
S. pneumonia, *H. influenza*, *M. pneumonia*, *Chlamydia*, etc.
Gram negatives, *M. tuberculosis*, *Legionella* species, other
Unknown pathogens

ARI (Williams et al 2002). In adults in the USA, there are an estimated 1 million cases of community acquired pneumonia each year and pneumonia is among the 10 most common causes of death. As noted in Table 1, there is wide variety of viral and bacterial pathogens commonly associated with ARI. Influenza viruses are the most important cause of serious viral ARI and respiratory syncytial virus (RSV) probably is the second most important viral respiratory pathogen. Influenza has most often been considered an important pathogen in adults but can also be a significant cause of ARI in children (Weinberg et al 2004). RSV is most often associated with serious disease in the infant and young child but also causes serious ARI disease throughout life (Falsey et al 2005). Our understanding of viral ARI is changing because of the discovery of novel viruses and the availability of better diagnostic assays. Five novel respiratory viruses including human metapneumovirus (van den Hoogen et al 2001), SARS coronavirus (CoV) (Drosten et al 2003, Ksiazek et al 2003, Peiris et al 2003), two novel human coronavirues, NL63 and HKU1 (Fouchier et al 2004, van der Hoek et al 2004, Woo et al 2005), and human bocavirus (Allander et al 2005) have been discovered since 2000. Improved diagnostics, especially sensitive polymerase chain reaction (PCR) assays, have also made it possible to consistently identify difficult to detect viruses such as rhinoviruses, and it is becoming increasingly clear that rhinoviruses have been under appreciated as serious respiratory pathogens (Miller et al 2007, Falsey et al 2002). Even with the discovery of novel viruses and improved diagnostics, there remain a significant number of ARIs for which the aetiology remains unknown. In adults, up to 50% of hospitalized patients with lower respiratory tract illnesses (LRIs) have no aetiological agent detected. Between 25% and 50% of children hospitalized with LRI have no aetiological agent detected. Some of these undiagnosed illnesses may be caused by as yet unknown pathogens and others by known pathogens but existing methods are either not sufficiently specific or not sufficiently sensitive to confirm the diagnosis. For example, detecting *Streptococcus pneumoniae* infection in the upper respiratory tract is not sufficiently specific to confirm it as the aetiology of pneumonia (Butler et al 2003). RSV infection is difficult to diagnose in adults because the assays traditionally used to detect infection are not sufficiently sensitive. In a study by Falsey et al (Table 2), most RSV infections (about 75%), diagnosed serologically with acute- and convalescent-phase serum specimens, are detected with a sensitive, nested PCR assay but only about one-third by virus isolation (Falsey et al 2002). Sensitive PCR assays can also improve our ability to detect viral respiratory infections in children as illustrated in a study by Weinberg et al (2004) (Fig. 1).

Novel viruses have been detected in illnesses of unknown aetiology, including ARIs, through various combinations of traditional and newer molecular methods.

TABLE 2 Detection of RSV in 1112 elderly patients

Serology	*Number*	*PCR+**	*Isolation +*
Positive	104 (9.4%)	74 (6.7%)	37 (3.3%)
Negative	1008 (90.6%)	13 (1.1%)**	6 (0.5%)

*Nested PCR assay.
**6/13 were isolation +.
Note: 117+ (serology detected 89%; PCR 74%; isolation 37%).
Adapted from Falsey et al (2002).

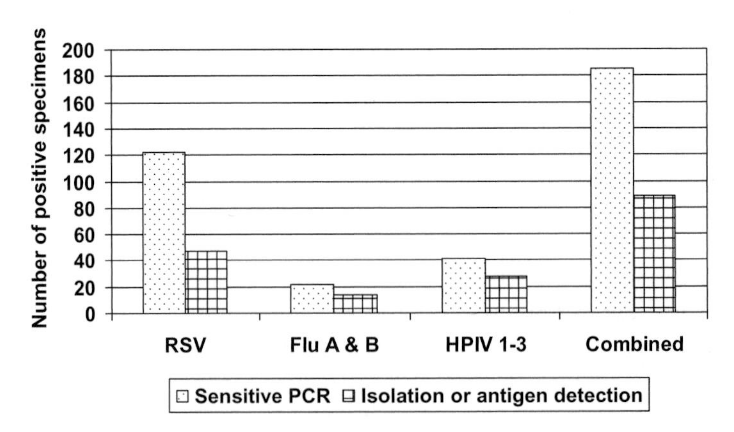

FIG. 1. Detection of respiratory pathogens: sensitive PCR versus isolation of antigen detection. Adapted from Weinberg et al (2004).

Steps often involved in successful efforts to identify novel viruses include: identifying diagnostic clues from clinical and epidemiological features of illness, collecting the right specimens, testing for known pathogens, amplifying the pathogen from specimens, purifying or selecting pathogen-specific material, detecting and then characterizing the pathogen, and determining the pathogen's role in disease. To date, tissue culture isolation has most often been the amplification step that provided sufficient quantity of novel virus for characterization studies but other methods such as random primer PCR amplification, representational difference analysis, and animal inoculation studies have also been successfully used. Once amplified, novel viruses have often been detected and characterized by combinations of techniques including electron microscopy, immunofluorescence or enzyme immunoassays with cross-reacting antisera, PCR assays using broadly reactive or random primers, sequencing studies, protein expression studies, etc. The Centers for Disease Control and Prevention (CDC) has investigated a number of outbreaks of unknown aetiology and two of these investigations, the outbreak of acute respiratory distress in New Mexico and the outbreak of severe acute respiratory syndrome (SARS), illustrate different ways novel pathogens have been identified.

In May 1993, the New Mexico Department of Health was notified of two persons from the same household who died of rapidly progressive respiratory distress syndrome (RDS) within 5 days. Surveillance for hospitalized cases or deaths from unexplained pulmonary oedema in the four corners area of Arizona, Colorado, New Mexico, and Utah identified 24 cases with a 50% mortality rate between late December 1992 and early June 1993 (Centers for Disease Control and Prevention 1993). Autopsy, urine, serum and respiratory specimens were submitted to the CDC and tested for known respiratory pathogens by PCR and serological antigen detection and isolation studies. The one positive result from these studies was the presence of Hantavirus-like IgG or IgM antibodies by enzyme immunoassay in some serum specimens from nine patients. Based on these serological results, respiratory and autopsy specimens were tested by PCR for sequences representing the two pairs of serotypes, the Hantaan and Seoul serotypes and the Puumala and Prospect Hill serotypes. The PCR for the Puumala and Prospect Hill serotypes amplified sequences from a novel Hantavirus designated as Sin Nombre virus (Nichol et al 1993).

In March of 2003, a global outbreak of SARS was noted and ultimately was associated with over 8000 cases and nearly 800 deaths. Within weeks of the outbreak's detection, several laboratories noted cytopathic effect (CPE) in tissue culture inoculated with specimens from SARS patients. At the CDC, electron microscopy demonstrated a coronavirus in the CPE-positive tissue culture and a PCR assay designed to detect any member of the coronavirus genera amplified novel coronavirus sequences. Two other laboratories identified a nearly identical virus by tissue culture and this virus, SARS CoV, was later proven to be the aetiologic agent of SARS (Drosten et al 2003, Ksiazek et al 2003, Peiris et al 2003).

A few other examples of successful pathogen discovery efforts further illustrate the variety of ways that novel viruses have been discovered. Hepatitis C virus was discovered when virus amplified by animal inoculation studies was purified by gradient centrifugation and its genome cloned for large scale peptide expression studies. Several of the expressed peptides reacted against convalescent serum specimens from ill patients and the sequences of the clones expressing these peptides provided the means to further amplify and characterize the hepatitis C virus (Choo et al 1989). The aetiology of Kaposi's sarcoma had eluded investigators until researchers used representational difference analysis to amplify sequences present in diseased but not in normal tissue; these sequences, in turn, provided the means to characterize the full genome of the previously unknown human herpes virus 8 (Chang et al 1994). Two groups used sequence independent methods to identify a novel coronavirus NL63 (Fouchier et al 2004, van der Hoek et al 2004). Both groups amplified viral sequences from CPE-positive tissue culture material inoculated with specimens from patients with ARI, characterized amplified products by gel electrophoresis and cloned bands not present in similarly amplified control tissue culture. Sequences from these clones were those of a novel coronavirus. In the final example, two novel parvoviruses were detected in large scale molecular screening studies that used Sequence Independent Single Primer Amplification (SISPA) and cloning methods to identify novel viral sequences. One of the viruses, human bocavirus (Allander et al 2005), was detected in specimens from patients with respiratory symptoms and has subsequently been associated with ARI. The other virus, PAR4, was detected in serum from patients with symptoms of an acute viral infection and has yet to be linked to human disease (Jones et al 2005).

At the CDC, we are developing several strategies to identify novel viruses from clinical specimens and/or tissue culture isolation material with each having different strengths and weaknesses. The initial strategy, pan viral family PCR assays, was developed to maximize sensitivity and reaction breadth for detecting novel virus. The pan viral family PCR assays were designed by using available sequences to identify the most conserved regions of the viral genome, usually the polymerase gene, and then using degenerate primers, inosine residues, and the CODEHOP technique (Rose et al 1998) to develop primers that should amplify all members of a viral family and/or genera within a family. To date PCR assays have been developed and evaluated for most of the viral families known to have members that cause encephalitis or respiratory disease (Table 3). These assays usually have had good sensitivity, i.e. usually <500 and often between 10 and 100 copies can be detected in the reaction mixture. For some viral families, such as those in the Paramyxoviridae, genera level primers had to be developed to achieve good sensitivity.

TABLE 3 PCR assays for viral families known to have members that cause encephalitis or respiratory disease

Virus family	*Viruses tested*	*Sensitivity of the PCR*
Adenoviridae	Representatives A–F	10–100 copies
Bornaviridae	BDV HE/80, NO/98	10–100 copies
Bunyaviridae (Bunyawera and CA gp)	CacheValley, LaCrosse	100–500 copies
Coronaviridae	All 3 antigenic groups	100–500 copies
Flaviviridae (Flavivirus gp)	JE, SLE, Den, WN, YF	100–500 copies
Herpesviridae	All human herpes viruses	10–100 copies
Orthomyxoviridae	A (all human and some avian), B, and C	10–100 for A or B; 50–100 for A–C
Paramyxoviridae	Representatives of all genera	10–100 for genera, 100–1000 for subfamily
Polyomaviridae	BK, SV40, JC	10–100 copies
Rhabdoviridae	VSV, HRV, IHVN, etc.	100–500 copies
Togaviridae (Alphaviruses)	WEEV, SFV VEEV	100–500 copies

TABLE 4 The sequence of steps that are required to establish a causal relationship between an agent and disease

Identifying diseases that might be caused by infection
- Detecting the agent in patients with disease
- Detecting the agent at the site of disease/symptoms

Establishing an association between infection and disease
- Detecting infection more often in ill than non-ill (control) patients

Establishing a causal link between infection and disease
- Replicating disease after challenge in an appropriate animal model and re-isolating/detecting the agent in the ill animal
- Detecting the pathogen at the site of disease/pathology by immune histology, *in situ* hybridization, or other techniques
- Detecting an immune response to the agent during illness
- Demonstrating a therapeutic response or protection from disease with agent-specific treatment or prevention

Identification of a novel virus is an important but initial step in studying a pathogen. A link between the pathogen and disease has to be established and Koch's postulates provide a standard for establishing a causal relationship. The initial postulates have been modified over the years to account for better methods to diagnose and study disease and a more complete understanding of disease pathogenesis. Table 4 outlines the sequence of steps that are required to establish a causal relationship between an agent and disease. The first step is to identify a disease that might be caused by the agent. The next step is to demonstrate epidemiologically that the agent is associated with disease, i.e. it is consistently detected more often in patients with disease than appropriate controls, and the final step is to demonstrate that the agent causes the disease. The classic way to establish a causal relationship is to reproduce the disease with the agent in an appropriate animal model. Other types of information, however, can help establish a causal relationship including demonstrating the pathogen by immune histology or *in situ* hybridization in diseased tissue, demonstrating an antibody response to the agent during the illness, and demonstrating a therapeutic response/decrease in disease with specific therapy or vaccination.

A causal link between the novel coronavirus, SARS CoV and the SARS disease was rapidly established. The WHO co-ordinated daily conference calls and a secure laboratory network website facilitated rapid information exchange and sharing of reagents and protocols. This rapid sharing was the key to quickly linking SARS CoV to disease. Within weeks of its discovery, data from most of the outbreak sites supported an association between SARS CoV and disease based on isolation, PCR and serological studies in cases and control populations (Ksiazek et al 2003, Peiris et al 2003). Within 2 months of discovery of the virus, researchers in the Netherlands had reproduced SARS pathology in a primate animal and fulfilled Koch's postulates for showing that SARS CoV caused SARS (Fouchier et al 2003).

Two new parvoviruses have recently been discovered. One human bocavirus (HBoV) has now been associated with ARI and the other PARV4 has yet to be associated with disease in humans. The fact that PARV4 has yet to be linked to disease illustrates the challenge that researchers can face in finding disease associated with a novel virus. Although the tools to do large scale screening for sequences from novel viruses are rapidly improving, methods to efficiently screen for possible disease associations have not improved in a parallel fashion. HBoV was detected during large scale molecular screening for new viruses in respiratory specimens from patients with ARI, therefore ARI was an obvious disease to consider for additional HBoV studies. Multiple groups have now demonstrated that HBoV is present in 2% to 10% of respiratory specimens from patients with ARI and was recently shown to be detected more commonly in ARI cases than controls (Table 5) (Fry et al 2007). It is, however, not yet clear if the virus causes ARI. Since a high percentage of bocavirus-positive specimens were also positive for other respiratory viruses, the role of HBoV in this disease is not clear (Table 6).

In summary, viral ARI is an important cause of human disease and much is yet to be learned about the agents, the clinical and epidemiological features of infection, and ways to treat and prevent the associated disease. Better diagnostic assays are changing our understanding of some of the previously characterized viruses and it is likely that investigators will continue to discover novel respiratory viruses. As tools for detecting novel viruses continue to improve, it is equally important

TABLE 5 HBoV infection in hospitalized pneumonia patients and controls, Sa Kaeo Province, Thailand

	Pneumonia cases		*Controls*	
Age group	*No. tested*	*% HBoV+*	*No. tested*	*% HBoV+*
1–11 mo	67	10	26	0
1–4 yrs	302	12	51	4
5–19 yrs	132	3	69	0
20–49 yrs	216	1	54	0
50–64 yrs	154	1	36	3
65+ yrs	307	1	36	0
Total	1178	4.5	280	1

Adapted from Fry et al (2007).

TABLE 6 HBoV infection and co-infection in hospitalized pneumonia patients, Sa Kaeo Province, Thailand

Age group	*No. tested*	*% Viral coinfection*
<1 yrs	224	93%
2–5 yrs	135	88%
6–18 yrs	95	100%
19–64 yrs	363	33%
65+ yrs	290	0
Total	1107	84%

Adapted from Fry et al (2007).

for investigators to develop strategies to efficiently screen for disease associations. One strategy that has proven very efficient is testing stored specimens from previous studies. When accompanied by good clinical and epidemiological data and control samples, these specimens will be invaluable to efforts to assess the role of a novel virus in human disease.

References

Allander T, Tammi MT, Eriksson M, Bjerkner A, Tiveljung-Lindell A, Andersson B 2005 Cloning of a human parvovirus by molecular screening of respiratory tract samples. Proc Natl Acad Sci USA 102:12891–12896

Butler JC, Bosshardt SC, Phelan M et al 2003 Classical and latent class analysis evaluation of sputum polymerase chain reaction and urine antigen testing for diagnosis of pneumococcal pneumonia in adults. J Infect Dis 187:1416–1423

Centers for Disease Control and Prevention 1993 Outbreak of acute illness—Southwestern United States. MMWR Morb Mortal Wkly Rep 42:421–424

Chang Y, Cesarman E, Pessin MS et al 1994 Identification of herpesvirus-like DNA sequences in AIDS-associated Kaposi's sarcoma. Science 266:1865–1869

Choo QL, Kuo G, Weiner AJ, Overby LR, Bradley DW, Houghton M 1989 Isolation of a cDNA clone derived from a blood-borne non-A, non-B viral hepatitis genome. Science 244:359–362

Drosten C, Gunther S, Preiser W et al 2003 Identification of a novel coronavirus in patients with severe acute respiratory syndrome. N Engl J Med 348:1967–1976

Falsey AR, Formica MA, Walsh EE 2002 Diagnosis of respiratory syncytial virus infection: comparison of reverse transcription-PCR to viral culture and serology in adults with respiratory illness. J Clin Microbiol 40:817–820

Falsey AR, Hennessey PA, Formica MA, Cox C, Walsh EE 2005 Respiratory syncytial virus infection in elderly and high-risk adults. N Engl J Med 352:1749–1759

Fouchier RA, Kuiken T, Schutten M et al 2003 Aetiology: Koch's postulates fulfilled for SARS virus. Nature 423:240

Fouchier RA, Hartwig NG, Bestebroer TM et al 2004 A previously undescribed coronavirus associated with respiratory disease in humans. Proc Natl Acad Sci USA 101:6212–6216

Fry AM, Lu X, Chittaganpitch M et al 2007 Human bocavirus: a novel parvovirus epidemiologically associated with pneumonia requiring hospitalization in Thailand. J Infect Dis 195:1038–1045

Jones MS, Kapoor A, Lukashov VV, Simmonds P, Hecht F, Delwart E 2005 New DNA viruses identified in patients with acute viral infection syndrome. J Virol 79:8230–8236

Ksiazek TG, Erdman DD, Goldsmith CS et al 2003 A novel coronavirus associated with Severe Acute Respiratory Syndrome. New Engl J Med 348:1953–1966

Miller EK, Lu X, Erdman DD et al 2007 Rhinovirus-associated hospitalizations in young children. Clin Infect Dis 195:773–781

Nichol ST, Spiropoulou CF, Morzunov S et al 1993 Genetic identification of a hantavirus associated with an outbreak of acute respiratory illness. Science 262:914–917

Peiris JS, Lai ST, Poon LL et al 2003 Coronavirus as a possible cause of severe acute respiratory syndrome. Lancet 361:1319–1325

Rose TM, Schultz ER, Henikoff JG, Pietrokovski S, McCallum CM, Henikoff S 1998 Consensus-degenerate hybrid oligonucleotide primers for amplification of distantly related sequences. Nucleic Acids Res 26:1628–1635

van der Hoek L, Pyrc K, Jebbink MF et al 2004 Identification of a new human coronavirus. Nat Med 10:368–373

van den Hoogen BG, de Jong JC, Groen J et al 2001 A newly discovered human pneumovirus isolated from young children with respiratory tract disease. Nat Med 7:719–724

Weinberg GA, Erdman DD, Edwards KM et al 2004 Superiority of reverse-transcription polymerase chain reaction to conventional viral culture in the diagnosis of acute respiratory tract infections in children. J Infect Dis 189:706–710

Williams BG, Gouws E, Boschi-Pinto C, Bryce J, Dye C 2002 Estimates of world-wide distribution of child deaths from acute respiratory infections. Lancet Infect Dis 2:25–32

Woo PC, Lau SK, Tsoi HW et al 2005 SARS coronavirus spike polypeptide DNA vaccine priming with recombinant spike polypeptide from Escherichia coli as booster induces high titer of neutralizing antibody against SARS coronavirus. Vaccine 23:4959–4968

DISCUSSION

Webster: There are potentially a vast number of, as yet, unidentified viruses in the plant and animal kingdom. Viruses at the human–animal interface are usually identified after a disease outbreak. How can we use the tools you describe to detect new viruses with zoonotic potential?

Anderson: It is one of the underlying themes: what do we do, what can we do, and what do we do with the data when we get them? What we have focused on, and what the tools available allow us to study, is looking at pathogens in the context of human disease. In the future, if the tools change and we have the right sets of specimens, then we could scan for new human pathogens in animals. Right now I don't think we have the tools or the resources to allow us to do this efficiently.

Osterhaus: If you want to go for new viruses in humans, why not make a proper inventory of what we have in animals? I think the way to go is to look for the gaps: those viruses found in related mammals that are not present or have not been found in humans. We should make an inventory of the holes or black spots. You have covered this with your systematic screen, where you have family based PCRs. This whole issue of syndromic surveillance, going for certain disease patterns in humans, could also be done the other way round. Take the viruses first and start looking for those viruses that have not yet been found in humans.

Anderson: Our panviral PCR strategy covers this within the families that we know cause human disease. We haven't gone back to the veterinary viruses which might show a different pattern. For example, there is a whole range of disease caused by parvoviruses in animals and it wouldn't be surprising if some of these cause similar diseases in humans.

Holmes: The difficulty of doing what you are saying is the prevalence issue. The number of animals you'd have to survey to pick up a RNA virus at a low prevalence is huge. Your chance of picking it up with no disease is limited. PCR wouldn't work. It is very hard to get blood samples from wild animals so you'd need to use fecal samples. It's hard to do in the absence of disease.

Osterhaus: I don't think so. There is a real opportunity: look at how many animals are being brought into the USA in quarantine, for example. You could do a systematic survey there.

Holmes: Look at the statistics of it. The animal is infected for three or four days and this is when you get virus. If the prevalence is say 5% the number you'd have to survey is enormous.

Osterhaus: You can do serology, too. I think there is too little emphasis on wild and domestic animals.

Holmes: I agree, but using universal primers isn't the way to do this. I would start with serology.

Osterhaus: I agree that serology is the way to go. Similar to the PCR screening, with serology there is a big challenge because we don't have all the serological assays for all the viruses we'd like to cover. We should start with the animals. Tackle the source. Rob Webster will agree with me when I talk about birds, and others will agree when I talk about bats and other animal species.

Webster: I agree with Ab Osterhaus, we should tackle the animal source. One example is the influenza gene pool in wild birds, another is the paramyxoviruses like Nipah and Hendra of bats that spread to pigs in Malaysia or horses in Australia.

Peiris: In bats, which are a important host for a number of viruses posing potential threats to human health including coronaviruses, we don't have good ways of doing serology because we lack a good second antibody reagent for immunoassays. Using bats as an example, PCR screening using conserved primers uncovers lots of viruses. It is likely (though not yet formally proved by mark–recapture type experiments) that coronaviruses establish persistent infections in bats. Some viruses of course may be shed for a short period and we could be looking for a needle in a haystack by going for virus detection.

Another relevant question in relation to screening for viruses in animal populations is how do we prioritize what we find? We might find many, many viruses and on what basis do we prioritise which ones we need to focus on in relation to zoonotic risk?

Holmes: For hepatitis C, people have been looking for an animal reservoir for this virus for years, and have been unsuccessful. There is nothing that looks like hepatitis C in any other animal. I do a lot of work with dengue, and our knowledge of sylvatic dengue is miniscule. We have very few isolates of sylvatic virus, yet I would bet good money that there are many more serotypes present in nature that could emerge again. The first thing to do is to change people's consciousness. We need to do more horizon scanning of biodiversity in nature, but people aren't funding this research.

Osterhaus: If you see how much money goes into upgrading hospitals when we have outbreaks, just a portion of this would make a big difference if we did some systematic screening by serology and PCR. In both domestic and wild animals it is important to do your syndromics surveillance: if you suddenly see mortality in wild species or certain disease signs spreading, this could be important.

Lai: Now we are detecting many new viruses because of the new techniques. But if we look at old data, there are many viruses that were isolated in the past but not characterized. For example, with coronavirus, people started to look for new coronaviruses in recent years after SARS and discovered NL63. Actually, in the early 1960s in the UK many coronaviruses were isolated using lung organ culture, and these were named OC1, 2, 3, 4 and so on. OC43 was studied as a prototype of coronaviruses because it grows well in culture. There are many human coronaviruses that have already been isolated, and many of them may be new coronaviruses, but we do not know because people haven't carefully characterized them. In someone's freezer somewhere there may be new viruses. If we have chances to characterise these, we may find that we already have more viruses than we realized.

Su: I was involved in the diagnosis of our SARS cases. But after reviewing the pathology slides, we ruled out bacterial infection and then we focused on the viral infection. I think the pathologist plays a very important role in providing the candidate pathogen in re-emerging infections. I don't know what happened to the cDNA array used in the identification of first SARS case by US-CDC. Has it been marketed?

Anderson: I assume you are referring to the multiple array strategy that was used to identify a coronavirus in tissue culture inoculated with specimens from SARS patients. This system is a very efficient way to identify a virus provided there is sufficient RNA/DNA in the sample. In the case of SARS, the multiple array worked on tissue culture isolation material but not on primary clinical specimens. My major concern with the multiple array for virus detection is sensitivity. My understanding is that it requires around 3000 to 30 000 copies of template while a sensitive PCR can often be designed to detect 10 or fewer copies of the template. For the respiratory viruses, we often need the most sensitive assays for their detection in clinical specimens. Consequently, the multiple array assays would not be our choice until their sensitivity has been improved.

Osterhaus: That is the general issue. We have to combine classical strategies and new strategies. If we can grow the virus this makes things easier.

Anderson: The multiple array technology is a nice way to look at tissue culture isolation, but for some clinical specimens it is not sufficiently sensitive.

Osterhaus: Unless you have diarrhoeal samples where there is enormous viral load, such as with rotaviruses.

Skehel: I'd like to ask about the bacterial contribution to the pathology. Is there anything known about the mechanism? Is it because of a secreted protease from the pneumococcus?

Anderson: My guess is that the virus predisposes to the bacterial infection rather than the other way around. There are some laboratory *in vitro* data showing that viruses can alter bacterial attachment and decrease local immunity and protection that may predispose to bacterial superinfection. Influenza virus infections are well recognized as predisposing to bacterial infection.

Skehel: Do you know about the prevalence of pneumococcus all the time in those sorts of patients?

Anderson: In young children, there can be a very high carriage of pneumococcus in the upper respiratory tract. Presumably RSV infection allows the bacteria to gain entry to the lower respiratory tract. In the study from South Africa there was a decrease in some categories of RSV LRI in vaccinated children compared to unvaccinated children, RSV pneumonia, compared to unimmunised children. There was also an increase in some of the other RSV-associated illness suggesting that pneumococcal vaccination protected from some RSV disease but may have increased the risk of other types of RSV disease.

Kahn: On the other hand, if you look at RSV bronchiolitis in children under two years of age, pneumococcal pneumonia as a complication is uncommon.

Osterhaus: But if you look at otitis media in young children, the primary infection is probably always the viral infection.

Peiris: You addressed the advantages in molecular technology which are helping us to discover new viruses. But we are still stuck with the same cell culture techniques of 30 years ago. The cells we use for virus isolation are no reflection of what is found in the respiratory tract, or whatever tissue we are interested in. Should we be looking at this question more and consider using primary cell cultures, or even differentiated primary cells for pathogen discovery?

Anderson: That's a good point. We may be able to find better tissue culture systems.

Osterhaus: The extreme of this would be using the appropriate animal model. If we use a monkey model we can reproduce the disease pretty well. The restriction here is that not all the human viruses will grow in primates, and in Europe we can no longer use higher primates.

Anderson: Some of the very early work on human coronaviruses included isolation studies in human volunteers.

Osterhaus: There may be some ethical stumbling blocks involved here!

The evolution of viral emergence

Edward C. Holmes

Center for Infectious Disease Dynamics, Department of Biology, Pennsylvania State University, University Park, PA 16802, USA and Fogarty International Center, National Institutes of Health, Bethesda, MD 20892, USA

Abstract. Despite the public health burden due to emerging viruses, little is known about the evolutionary processes that allow viruses to jump species barriers and establish productive infections in new hosts. Understanding the evolutionary basis of virus emergence, and whether generalities exist in this process, is therefore a key research goal in the study of infectious disease. Herein, I discuss the evolutionary biology of viral emergence, set within the analytical framework of fitness landscapes provided by population genetics, and the possible reasons why some viruses may be more likely to emerge than others. Particular emphasis is given to the complex interplay between rates of adaptive evolution, the ability to recognize cell receptors from host species with different levels of phylogenetic divergence, and the likelihood of cross-species transmission. Alarmingly, there is still a lack of definitive data on many key aspects of viral emergence, particularly whether this process routinely requires adaptation to the new host species during the early stages of infection, or is in part a chance process involving the transmission of a viral strain with the necessary genetic and phenotypic characteristics. Systematic studies of genetic and phenotypic diversity within individual hosts, a critical aspect of viral fitness, are particularly notable for their absence.

2008 Novel and re-emerging respiratory viral diseases. Wiley, Chichester (Novartis Foundation Symposium 290) p 17–31

The nature of emerging viruses

Emerging viruses have been a long-standing threat to public health. Since its first description in the early 1980s, the human immunodeficiency virus (HIV), the causative agent of AIDS, has infected hundreds of millions of people, a large proportion of whom have died. Despite our increasing understanding of this virus, around 4.1 million new infections occurred in 2005, 70% of these in sub-Saharan Africa (*http://www.unaids.org/en*). As an emerging virus HIV is not alone in its severe consequences for human morbidity and mortality. For example, while only 8437 people (with 813 deaths) were infected by the SARS coronavirus in 2002–2003, its global economic cost has been estimated in excess of $50 billion. Finally, although at the time of writing there have been only 277 documented human infections with H5N1 avian influenza virus, resulting in 168 deaths, the toll of human disease could be enormous if the virus manages to evolve human-to-human transmission and appropriate intervention is not forthcoming (Ferguson et al 2005).

Given the past, present and potential for emerging viruses to adversely affect human health it is clearly essential to determine the factors that enable viruses to enter and spread through new host populations, and why some viruses seem better able to emerge than others. This information may ultimately allow us to predict, at least in part, what types of virus may emergence in human population in the future, where such emergence is likely to occur, and what animal species are most likely to act as reservoirs.

At present, few generalities can be drawn about the evolution of viral emergence. The most transparent is that emerging viruses usually have their ancestry in related viruses that infect other animal species, so that viral emergence usually equates to cross-species transmission (Cleaveland et al 2001). For example, the origins of HIV clearly lie in the closely related viruses that infect chimpanzees (SIVcpz) in Central-West Africa (Hahn et al 2000). Similarly, although the ultimate species reservoir of SARS coronavirus is still unclear, related coronaviruses are commonly found in horseshoe bats (Li et al 2005).

If cross-species transmission is the key to viral emergence, and because humans interact closely with many animal species, then it must be that we are continually exposed to new viral pathogens. However, very few of these eventually establish themselves in human populations. Consequently, at each exposure event there is only a very small chance that the foreign virus will take hold. As such, we can define three types, or stages, of an emerging infection: (1) those that 'spill-over' from a donor to a recipient host species with no onward transmission, (2) those that result in only local chains of transmission ('outbreaks') in the recipient host species, and (3) those that develop into a full-blown epidemic with sustained onward transmission in the new host species. A key issue in studies of viral emergence is therefore to identify the factors that determine whether an initial infection will survive in a new host species and develop into a fully fledged epidemic.

The population genetics of viral emergence

Until recently, most studies of viral emergence concentrated on the ecological factors that underpin cross-species transmission and epidemic spread. Such studies gave particular emphasis to how changes in human ecology, notably increases in population size, changes in land use and global travel, have been responsible for an elevated burden of infectious disease by increasing the proximity and/or density of host and/or reservoir populations (Morse 1995). In general, this work demonstrates that successful emergence depends on two key ecological variables: the contact network of the recipient population and the likelihood of individuals within this population transmitting the virus. The number of secondary cases produced when a new virus is introduced into an entirely susceptible population is given by the basic reproduction number, R_0. As individuals become infected, then recover or die, the proportion of susceptibles declines along with the number of secondary cases per infection, R. If $R < 1$, as is the case for most spill-over infections (stage 1 in the emergence process described above), an infection will not cause a major epidemic. However, when $R > 1$ a new infection may spread locally (emergence stage 2), with potential for further, more widespread epidemic transmission (emergence stage 3) (Kuiken et al 2006). However, such ecological studies are not aimed at identifying the common *evolutionary processes* that determine why R is <1 in most cases and >1 in a few. Indeed, explaining the spectrum of epidemic outcomes following exposure to a novel pathogen is perhaps the biggest challenge facing those studying emerging viruses. Consequently, it is critical to determine whether there are any general evolutionary rules governing the transition from spill-over to full-blown epidemic. In general terms, two models can be used to explain the evolution of viral emergence:

Model 1: adaptation in the new host species is the key parameter in viral emergence

In most discussions of viral emergence it is assumed that viruses have to adapt to successfully spread in a new species because, at the time of initial exposure, they lack many of the mutations that facilitate onward transmission (and human-to-human transmission would be selectively favoured as it increases R). Adaptive evolution is therefore the process that takes viruses from epidemic stages 1 through 3 as it elevates R to >1, therein allowing the establishment of longer chains of transmission (Antia et al 2003). This adaptive process is thought to occur during the short chains of inter-host transmission that might characterize the early stages of an epidemic (Antia et al 2003). Hence, those human viruses that have not developed sustained human-to-human transmission are simply those that have not fully adapted to our species ($R < 1$). Different host species can therefore be thought as representing different fitness peaks on an 'adaptive landscape' of viral genomes, and in most cases traversing between these peaks—the process of cross-species transmission—is difficult because they are connected by valleys of low fitness (Fig. 1). The more mutations required for a virus to move between peaks, the deeper the valley and the less likely that it can be crossed in a single step. As a case in point, direct adaptation seems to have been central to the emergence of dengue virus in humans (Moncayo et al 2004), of the canine parvoviruses (Shackelton et al 2005), and perhaps also of HIV (Schindler et al 2006).

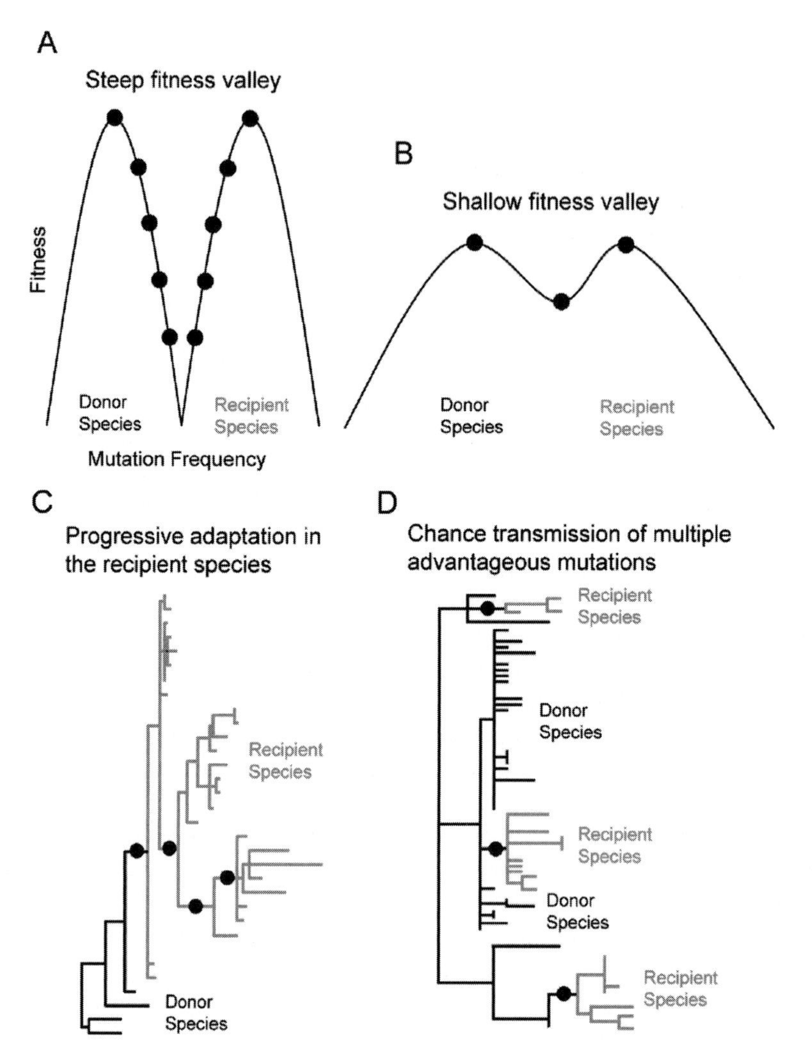

FIG. 1. Contrasting evolutionary models for the cross-species transmission of viruses. (A) The donor and recipient species represent two fitness peaks separated by a steep fitness valley. Multiple advantageous mutations (closed circles) are therefore required for the virus to successfully establish onward transmission in the recipient host species. (B) The donor and recipient species are separated by a shallower fitness valley. This facilitates successful cross-species transmission as a smaller number of advantageous mutations are required. (C) When multiple mutations are needed for a virus to adapt to a new host these may evolve progressively in the recipient species. However, this necessarily requires some onward transmission. (D) It is also possible that many of the mutations required to allow adaptation to a new host species were pre-existing in the donor species and transmitted simultaneously. This will accelerate successful viral emergence. (Figure taken from Kuiken et al 2006 Science 312:394–397.)

Model 2: exposure is the key parameter in cross-species virus transmission

A second model of viral emergence proposes that cross-species transmission simply requires the chance exposure of a virus to a new susceptible population, with little adaptive evolution, as the virus in the donor already possesses many of the mutations required to replicate and transmit in a new host. Under this model, the critical parameter determining if successful onward transmission will occur is the probability that the recipient population is exposed to a viral strain that, by chance, already harbours some of the mutations required for successful onward transmission. The successful emergent strains are therefore those that by chance are *pre-adapted* to cause productive infections in the new host species, so that the probability of emergence then becomes a function of the frequency of exposure. Of the multitude of RNA viruses produced by erroneous replication within an individual host, some, by chance, may possess those mutations that allow productive spread in a new host. Indeed, that the majority of emerging infections result in dead-end infections implies that even short-term transmission chains are difficult to establish for most viruses because they lack the necessary mutations. Further, for most emergent viruses it has been difficult to establish that cross-species transmission is associated with direct adaptation. For example, although some studies suggest that SARS-CoV was subject to adaptive evolution during its early spread through humans (Yeh et al 2004), it is unclear whether this reflects adaptation to the new host species or selection for immune escape after sustained transmission had been established.

Currently, arguments can be advanced for both models with little empirical data to choose between them. The debate over the potential emergence of H5N1 influenza A virus in humans provides an illustrative example. One factor that is essential (although not sufficient) in determining whether avian influenza viruses are able to cause human epidemics are the sialic acid cell receptors found on cell-surface oligosaccharides. All avian influenza viruses replicate in the gastro-intestinal tract and bind to sialic acid in an α2,3-linkage to galactose, while human influenza viruses replicate in the respiratory tract and bind to sialic acid in an α2,6-linkage. Hence, the shift between sialic acid linkage is critical in enabling the switch from birds to humans and often involves changes at two amino acid residues in the viral haemagluttinin (HA). The key evolutionary question is therefore whether these mutations appear *de novo* in humans, in short transmission networks of people who initially suffer avian influenza, or if they pre-exist in the avian population so that emergence will occur if the appropriate strain is transmitted? Despite their undoubted importance, there is currently little data to determine the applicability of these competing models of viral emergence, particularly as intra-host genetic diversity has rarely been examined in viruses causing acute infections, including influenza A virus.

To determine which of these models of viral emergence best fits natural situations it is essential to conduct more intricate studies of viral genetic variation in both their donor and recipient species. In particular, if cross-species transmission routinely requires adaptive evolution, then the ability to jump species boundaries should be directly related to the extent and structure of genetic variation within a viral population. This would then explain why most emerging viruses are rapidly evolving RNA viruses and retroviruses (Cleaveland et al 2001, Woolhouse et al 2001, Woolhouse 2002). In contrast, DNA viruses are more often associated with a process of virus–host co-divergence that can extend many millions of years, perhaps because they often establish persistent infections and so can more easily track host evolution (Holmes 2004). Rates of nucleotide substitution for RNA viruses are usually in the realm of 10^{-3} to 10^{-4} substitutions per site, per year (subs/site/year), whereas in DNA viruses rates of 10^{-7} to 10^{-8} subs/site/year are usually more typical (Moya et al 2004). This elevated capacity for evolutionary change obviously allows RNA viruses to quickly generate the mutations that might be required to adapt them to new host species. The relationship between genetic diversity and emergibility further suggests that the closer the phylogenetic relationship between the donor and recipient species—for example, humans and other primates—the more likely that the virus will already possess sufficient mutations to allow cross-species transmission. In contrast, the more phylogenetically divergent the donor and recipient species, such as humans and avian species, then the more novel mutations are likely to be required to adapt the virus to new hosts (corresponding to the steep fitness peaks in Fig. 1A).

Unfortunately, there are a variety of other factors that cloud the potentially simple relationship between high rates of evolutionary change and the ability to jump species boundaries. First, many aspects of the viral life-cycle need to be optimized to allow successful transmission among members of a new species. An important step for many viruses is the virus–host cell receptor interaction (Baranowski et al 2001). Because all viruses have an absolute dependence on the cellular machinery of their host for successful replication, the interactions between viral proteins and host cellular receptors constitute an important part of the viruses' fitness landscape, and is therefore an arena where selection will be potent; purifying selection is expected to act quickly against those strains that lack the mutations needed for receptor recognition. It has also been suggested that 'generalist' viruses, that infect a broad range of cellular receptors, may be more able to cross species boundaries than 'specialist' viruses that have a narrower tropism and therefore need to infect hosts that harbour the same cell types. Provisional analyses suggest that this is indeed the case, as viruses that utilize conserved cell receptors are more able to jump species boundaries than viruses that use genetically divergent cell receptors (Woolhouse 2002).

Second, while mutation clearly provides the raw materials for evolutionary change, it is the dual processes of genetic drift and natural selection that determine how viruses might respond to the challenges of replication in new hosts. As such, the rate of virus adaptation is not only determined by the overall rate at which mutations arise, but also by the fitness of these mutations, particularly the proportion that are advantageous in multiple hosts. Notably, although RNA viruses often show extremely high levels of genetic variation, it is likely that a large proportion of the mutations that arise within an individual host are deleterious (Elena & Moya 1999, Sanjuan et al 2004a), so that only a small proportion of the total mutational spectrum will increase fitness. Similarly, there is mounting evidence that mutations often show complex epistatic interactions, which can have major effects on the rate of adaptation (Sanjuan et al 2004b). Also, because an emergent virus will only infect one or a small number of individuals when it first enters a new population, genetic drift is likely to play a major role in determining what viral mutations are fixed in the early phases of emergence. The respective influences of drift and selection will be dependent upon the size of the population bottleneck that accompanies viral transmission among hosts, itself a function of the mode of transmission, and whether this bottleneck is random, reflecting a general reduction in viral population size, or selective, so that the loss of genetic diversity corresponds to the outgrowth of the fittest variant. However, there are currently few data for any virus on either the magnitude of the transmission bottleneck or whether it is neutral or selective in nature.

Third, it is now clear that there are important constraints in viral evolution, many of which will undoubtedly influence the process of emergence. The most obvious of these relate to mode of transmission. In particular, many emerging viruses are transmitted by arthropod vectors, which could potentially take blood meals from a range of mammalian hosts thereby facilitating cross-species transmission. However, there is also an association between transmission mode and the ability of a virus to successfully replicate in the cells of a new host species; for example, both experimental and comparative studies have revealed that arboviruses are constrained compared to viruses transmitted by other mechanisms (Holmes 2003, Woelk & Holmes 2002, Zárate & Novella 2004). This effect is most likely due to antagonistic pleiotropy; mutations that increase fitness in one host species reduce it in another and so move from the advantageous to deleterious class (and the greater the phylogenetic difference between these species, the greater these fitness trade-offs are likely to be). More compelling is that although arboviruses are frequently associated with sporadic disease in humans, few are able to sustain long-term transmission networks (Woolhouse, personal communication). Hence, the intricate adaptive solutions required to replicate in divergent hosts, and the increase in the proportion of deleterious mutations this entails, may act to prevent many arboviruses from evolving sustained transmission networks in new host species by reducing the effectiveness of adaptive evolution. A fundamental challenge for viruses is therefore that they require mutations that by adapting them to recipient host species, reduce their fitness in the donor host. The nature of fitness trade-offs, and how they affect the process of cross-species transmission, is therefore one of the most important questions in the study of emerging viruses.

Lastly, although high mutation rates are the engine of evolutionary change in RNA viruses, there is mounting evidence that genetic variability can be shaped, at least in part, by recombination (or reassortment). Because recombination can potentially increase fitness by creating advantageous genotypes and removing deleterious ones, it might also be supposed that it plays a vital role in viral emergence. In particular, recombination and reassortment will allow some viruses to acquire many of the mutations necessary for host adaptation in a single evolutionary step, and hence make a major jump in fitness space. Similarly, the purging of deleterious mutations by recombination may free advantageous mutations from their low fitness baggage, therein increasing the overall rate of adaptation. The evolutionary potential of recombination is perhaps best documented in the primate lentiviruses, such as HIV, which not only experience extremely high rates of recombination (Jung et al 2002), but where this process has been directly associated with cross-species transmission (Bailes et al 2003). Similarly, the cross-species transmission of influenza A virus from birds to mammals is often associated with reassortment among HA and neuraminidase (NA) subtypes as well as other gene segments (Horimoto & Kawaoka 2005, Webby & Webster 2001).

However, while recombination may allow viruses to traverse the adaptive landscape faster than through mutation alone, the optimal epistatic interactions among genes are likely to be broken by recombination, and most recombinants, like most point mutations, are also expected to be deleterious. For example, as the influenza A virus HA and NA proteins both act on the sialic acid receptors on cells they may need to acquire complementary changes to enable the optimal use of the receptors in the new host species. Another reason to doubt the role played by recombination in viral emergence is that, other than in the retroviruses, recombination may not be a particularly common process in RNA viruses. Recombination appears to occur particularly sparingly in negative-sense RNA viruses (Chare et al 2003), most likely because their RNA is always encapsidated, which limits the rate of template-switching. As a number of emerging viruses have negative-sense RNA genomes, this argues against recombination as a general process in viral emergence, although, of course, rare events like those generated by recombination may sometimes be critical in kick-starting the process of viral emergence. Overall, most of the available evidence suggests that recombination rates in RNA viruses are controlled by two factors; the ability of the virus in question to undergo template switching and the frequency with which multiple infections occur.

However, like so many elements of the study of viral emergence, in reality there are few data to choose among competing theories.

For the study of the evolution of viral emergence to come of age a number of key issues need to be addressed. Most notably, perhaps, a broader knowledge of the genetic and phenotypic diversity of viral populations sampled from *within* individual hosts is urgently required. Understandably, the study of acute RNA viral infections, which represent the most common type of emerging viruses, has traditionally focused on the study of single consensus sequences. By necessity, these sequences describe the *average diversity* of the intra-host viral population, masking a myriad of variable mutant sequences. Although the analysis of such consensus sequences is of great value for many aspects of molecular epidemiology, such as inferring the patterns and process of epidemic spread through populations, it is of less utility for the study of viral fitness landscapes that are essential to understand how RNA viruses are able, or not, to successfully jump species barriers. The development of rapid, high-throughput, methods of DNA sequencing may therefore herald a new age in the study of both evolutionary biology and disease emergence.

References

Antia R, Regoes RR, Koella JC, Bergstrom CT 2003 The role of evolution in the emergence of infectious diseases. Nature 426:658–661

Bailes E, Gao F, Bibollet-Ruche F et al 2003 Hybrid origin of SIV in chimpanzees. Science 300:1713

Baranowski E, Ruiz-Jarabo CM, Domingo E 2001 Evolution of cell recognition by viruses. Science 292:1102–1105

Chare ER, Gould EA, Holmes EC 2003 Phylogenetic analysis reveals a low rate of homologous recombination in negative-sense RNA viruses. J Gen Virol 84:2691–2703

Cleaveland S, Laurenson MK, Taylor LH 2001 Diseases of humans and their domestic mammals: pathogen characteristics, host range and the risk of emergence. Phil Trans R Soc Lond B Biol Sci 356:991–999

Elena SF, Moya A 1999 Rate of deleterious mutation and the distribution of its effects on fitness in vesicular stomatitis virus. J Evol Biol 12:1078–1088

Ferguson NM, Cummings DA, Cauchemez S et al 2005 Strategies for containing an emerging influenza pandemic in Southeast Asia. Nature 437:209–214

Hahn BH, Shaw GM, de Cock KM, Sharp PM 2000 AIDS as a zoonosis: scientific and public health implications. Science 287:607–614

Holmes EC 2003 Patterns of intra- and inter-host nonsynonymous variation reveal strong purifying selection in dengue virus. J Virol 77:11296–11298

Holmes EC 2004 The phylogeography of human viruses. Mol Ecol 13:745–756

Horimoto T, Kawaoka Y 2005 Influenza: lessons from past pandemics, warnings from current incidents. Nat Rev Microbiol 3:591–600

Jung A, Maier R, Vartanian JP et al 2002 Recombination—multiply infected spleen cells in HIV patients. Nature 418:144

Kuiken T, Holmes EC, McCauley J, Rimmelzwaan GF, Williams CS, Grenfell BT 2006 Host species barriers to influenza virus infections. Science 312:394–397

Li W, Shi Z, Yu M et al 2005 Bats are natural reservoirs of SARS-like coronaviruses. Science 310:676–679

Moncayo AC, Fernandez Z, Ortiz D et al 2004 Dengue emergence and adaptation to peridomestic mosquitoes. Emerg Infect Dis 10:1790–1796

Morse SS 1995 Factors in the emergence of infectious diseases. Emerg Infect Dis 1:7–15

Moya A, Holmes EC, González-Candelas F 2004 The population genetics and evolutionary epidemiology of RNA viruses. Nat Rev Microbiol 2:279–287

Sanjuan R, Moya A, Elena SF 2004a The distribution of fitness effects caused by single-nucleotide substitutions in an RNA virus. Proc Natl Acad Sci USA 101:8396–8401

Sanjuan R, Moya A, Elena SF 2004b The contribution of epistasis to the architecture of fitness in an RNA virus. Proc Natl Acad Sci USA 101:15376–15379

Schindler M, Münch J, Kutsch O et al 2006 Nef-mediated suppression of T-cell activation was lost in a lentiviral lineage that gave rise to HIV-1. Cell 125:1055–1067

Shackelton LA, Parrish CR, Truyen U, Holmes EC 2005 High rate of viral evolution associated with the emergence of canine parvoviruses. Proc Natl Acad Sci USA 102:379–384

Webby RJ, Webster RG 2001 Emergence of influenza A viruses. Phil Trans R Soc Lond B Biol Sci 356:1817–1828

Woelk CH, Holmes EC 2002 Reduced positive selection in vector-borne RNA viruses. Mol Biol Evol 19:2333–2336

Woolhouse MEJ 2002 Population biology of emerging and re-emerging pathogens. Trends Microbiol 10:S3–S7

Woolhouse MEJ, Taylor LH, Haydon DT 2001 Population biology of multihost pathogens. Science 292:1109–1112

Yeh S-H, Wang H-Y, Tsai C-Y et al The National Taiwan University SARS Research Team 2004 Characterization of severe acute respiratory syndrome coronavirus genomes in Taiwan: molecular epidemiology and genome evolution. Proc Natl Acad Sci USA 101:2542–2547

Zárate S, Novella IS 2004 Vesicular stomatitis virus evolution during alternation between persistent infection in insect cells and acute infection in mammalian cells is dominated by the persistence phase. J Virol 78:12236–12242

DISCUSSION

Webster: Let me ask you a question about H5N1: it has been around for 10 years now. If we use the rules you have been putting forward, can we relax?

Holmes: All I would say is that the H5N1 virus in 2007 is no more adapted to humans than it was in 1997. It is evolving in avian species and occasionally spills over into humans. It is not progressively picking up mutations that make it human adaptive. It doesn't mean it won't happen. To speculate about what is going to happen is foolish.

Vasudevan: Is that because there are not enough numbers?

Holmes: No, it's because evolution is about rare events. It would be foolish to say it is not going to happen because it hasn't happened for 10 years. We can't make a probability distribution for it because we haven't got enough information.

Osterhaus: It might be happening in the birds themselves. In the Netherlands, in 2003 one person died of a H7N7 virus that already had a number of mutations.

You pointed out that these viruses that are outliers represent evolutionary dead ends. They may be very important for intraspecies transmission.

Holmes: There is too much time and energy wasted on trying to predict what is going to happen. The critical thing is great surveillance. If I had a request as a user, it would be that we should look at intrahost genetic diversity in avian species more frequently.

Peiris: When you make the statement that the H5N1 virus now is no more adapted to humans than in 1997, do we know what mutations are required to make it more adapted?

Holmes: No.

Webster: We might want to go round the table and challenge the statement that it is no more adapted now.

Holmes: I have looked at the mutations that are supposedly definitive human ones and definitive avian ones. We see some of them coming up in a human transmission cluster, but each successive bird we look at doesn't have them again. These mutations are not selectively advantageous in birds.

Osterhaus: Hasn't the pathogenicity for mammals in general changed since 1997?

Holmes: I'm not the expert on this.

Su: With SARS coronavirus, I recall that there was a 29 nucleotide sequence that led to the jump from animal to human.

Holmes: I haven't looked at SARS. We can think of evolution in a very simple way of 'will it or won't it adapt', but we need to move beyond this. The parameters that shape what selection favours are more intricate than we might think. We are missing some of the critical data.

Peiris: You made an interesting point about the constraints between mutations and genome size, with a trade-off between the two. What about a virus with a segmented genome? Does this confer any difference to this?

Holmes: Segmented viruses are a bit longer but not hugely longer. They don't break the rule. There appears to be a limit in RNA viruses for ORF size. We don't see ORFs longer than 3000–4000 amino acids. Why are viruses segmented? I suspect segmentation is about controlling gene expression.

Osterhaus: Reassortment, too. This gives them versatility.

Holmes: If you are a segmented RNA virus you carry the same deleterious mutation load as a non-segmented RNA virus. One of the classic theories of reassortment in evolution is that it is there as a way of purging bad mutations. This doesn't appear to work according to our data. Segmentation is about controlling gene expression and reassortment is a side-effect of this.

Osterhaus: Look at flu. It is a virus that can sustain itself in different animal species also because it reassorts.

Holmes: But the reassortment mostly breaks up good associations. This is the big evolution of sex question. As well as creating good things, reassortment also breaks up good things. This is the problem.

Smith: I'm interested in what you are saying about the internal nodes on the tree, and that the tip nodes are all deleterious. Where are the viruses that don't have deleterious mutations? They must be on the tip, also.

Holmes: I think there is lots of complementation going on. Much of what is sequenced is deleterious. We see this in the NY state influenza data. For example, I have a big deletion in the neuraminidase, which is not a clone, but a consensus sequence. This is in a bit of neuraminidase that never varies, yet this deletion is there at high frequency. This must have been complemented. How RNA viruses survive their high error rate is to have a massive progeny number. This is their life strategy: a big population allows them to survive a high mutation rate.

Smith: We know from experiments that we are involved with in horses that there are a lot of deleterious mutations, but the consensus will blank those out, averaging things out. The consensus is not on average going to be a deleterious mutation. Those consensus sequences are the sequences at the tip of the tree.

Holmes: Yes, but some of them also contain deleterious mutations. Selection will take a while to purge deleterious mutations. What is sequenced in an individual host may be OK in that host, but in a new host it becomes deleterious since fitness landscapes change so quickly. It can take a while for selection to purge something out.

Smith: In the times when pathogenicity is related to increased viral replication, for example the T249B substitution, because of the replication rate of the virus this substitution is being produced frequently. If it is purely a matter of replication, how come the T249B high replication rate mutant does not quickly become the dominant strain?

Holmes: That's virulent in American crows. I don't know what would happen if it is put in an old world crow species. Fitness is context dependent. It depends on what strain of host species you are in. You can't generalize what is seen in American crows to what is seen in British starlings.

Lane: In a virus infection of this type, with a high frequency of the virus, how many cells are simultaneously infected with two different viruses with two different sequences? With SV40 we have this problem of what we call inactive passenger virus. The virus always has to be complemented. When we isolate viruses from natural populations with DNA viruses, we always find that they are a mixture of replication competent and replication incompetent viruses. This represents a bigger gene pool because we have material in which replication and recombination can take place, where you are actually carrying a larger density of variation as a population. The implication of what you said is that the individual host is holding a pool of varying viruses, and this is the infectious agent, not a single virus. Is that the case?

Holmes: Absolutely. The mutational dynamics guarantee that. When you are infected by a virus, it is not one particle, but a population. Mutation rates are so high and turnover is high, by definition you will have a population of viruses.

Lane: You have portrayed this virus as being right on the edge of being alive because the mutation load is so high. Is this why nucleosides are so effective as anti-retroviral therapies?

Holmes: Yes, there is a whole school of drug therapy based on this idea of taking the error rate over the error threshold. It works, at least *in vitro.*

Osterhaus: I'd like to challenge the dogma you started with about RNA viruses. being the cause of newly emerging infections. There are several examples of small (e.g. parvo-) and large DNA viruses. One example is Aujeszky's disease which is a porcine virus that may cause an outbreak in cattle.

Holmes: I said 'most', not 'all'.

Lai: You put the mutation rates for the DNA viruses in one range, and for RNA viruses in another. Is it true that in each category the mutation rates are the same?

Holmes: No. In general it is true. What is happening with the evolutionary biology of viruses is that the distinction between RNA and DNA viruses is becoming greyer. There are a few cases where some RNA viruses are evolving very slowly. An example would be the simian foamy viruses where the replication rate is low, even though the error rate is probably normal. There are a couple of cases of small DNA viruses where single stranded ones are as fast as RNA viruses.

Lai: Classic studies by Peter Palese 20 years ago showed that when he compared polioviruses with flu viruses, the flu viruses had error rates 10 times higher.

Holmes: The range of error rates in those that use RNA polymerise are all within a log of each other. Retroviruses are lower: their transcriptase has a slightly higher fidelity, and the error rate is about a fifth of the rate of RNA polymerise. But they are all way higher then DNA polymerises. A log is not much.

Kahn: One comment about the overlap between fitness and pathogenesis: it seems to me that fitness is the ability of the virus to replicate efficiently in a host. This doesn't necessarily correlate with pathogenesis. You gave the example of the West Nile virus proline substitution. Is there evidence that this proline substitution actually increases the fitness?

Holmes: This site was the only one I found selected in the whole West Nile virus genome. It is increasing in frequency because it is advantageous.

Kahn: For many of the viruses we are talking about, humans are dead end hosts. Fitness may not be an issue for these viruses. How do you account for this in these viruses?

Holmes: This is a big question in evolution: the evolution of virulence, and it is a complicated topic. The general idea is that there is a trade-off between virulence and transmissibility. There is an optimum virulence that enhances the transmissibility of the virus. You don't want to kill the host too quickly, but there needs to be enough virus to infect a new host. The West Nile virus is interesting because the American crows die quickly. I can't understand why this is advantageous. My theory is that American crows are not the main reservoir species. Mutation is favoured in the reservoir species because it aids transmission, but a spin-off from this is that it kills the crows.

Osterhaus: Plus you will have selection in the crows. The American crows may eventually become like European crows. Your definition of fitness there regards the individual, but a proper definition of fitness should involve the whole population and should include transmissibility as well.

Anderson: Early on in SARS there was evidence of adaptation in the S protein based on rate of change in the S gene. Later in the outbreak, the rate of change in the S gene decreased. Some of these changes probably improved its ability to replicate in human cells and possibly its transmissibility. Does this mean a virus might reach some point in adapting to a new host when sequential mutations do not work but in some instances complementary mutations, which individually may be deleterious, are needed to move to the next level in improving its replication and/or transmissibility in a host. This could create a bottleneck and a perceived inability of the virus to continue adapting to a host. I wonder whether this is the case with H5N1? If sequential changes would lead to efficient replication and transmission in humans, would it have occurred in the last 10 years by chance?

Holmes: That's the sharp fitness landscape. I don't think SARS did this very well. Do we know that the adaptation seen early on in SARS wasn't immune evasion rather than host adaptation? I don't know how you can distinguish them.

Anderson: For SARS coronavirus, *in vitro* studies suggest it adapted to the binding to the human ACE2 receptor.

Holmes: In general, you are right. Compensating mutations are a real limiting step. You make every single mutation every few hours. Every double mutation of a particular pair takes far longer to get statistically. HIV CTL (cytotoxic T lymphocyte) escape is a good example of this. There are some CTL escape mutations that take many years to evolve. This is because individually this mutation destroys the fitness of the capsid protein. For it to work you need two others to occur to restore this, and this needs to happen in the same mutation cycle. Even with the mutational power of RNA viruses this is a lot harder to do. We need to think about this epistatically; it is not simple additive fitnesses.

Smith: Let's go back to the 249 substitution and the hypothesis that the virus is going to adapt to American crows such that it doesn't kill them quickly. We know that this substitution will occur frequently. Is it more the case, then, that the adaptation is not that this substitution will not be seen, but that there would be adaptations elsewhere in the genome that would make the substitution disadvantageous for some structural reason, so that if it does happen the virus is no longer going to be highly pathogenic or, potentially, not even viable.

Holmes: I agree. We need to think about the interactions between the genes and the sites. I don't think we know this for any real system yet.

Smith: To look for positive selection, one would have to look everywhere but that site.

Holmes: In that case I looked carefully, and this was the only mutation that has that pattern. All the other old world West Nile viruses also had the proline at this position. It is the only region in the whole genome that has that pattern. In general, to understand fitness we can't just think of one mutation, but rather of all the epistatic interactions.

Peiris: You mentioned that in terms of interspecies transmission, moving from a primate to a human is the easiest thing to do. But when we look at most of the emerging viral infections, very few have come from primates.

Holmes: I was talking about the genetics of adaptation. Genetically what I am saying is correct. We also have the ecology to consider, which is about exposure. We aren't exposed to primates and their viruses.

Osterhaus: You talked about the host: the American crow and humans. You more or less take for granted that each is a homogeneous species. In reality, there are super-spreaders in the population. There is genetic diversity on the part of the host.

Holmes: You are correct: there are many places with dengue where we can see a particular strain of *Aedes aegypti* has allowed the virus to emerge. In northern Mexico there is a strain of this mosquito that doesn't carry dengue, but in southern Mexico it does.

Osterhaus: There is also HLA diversity that may be largely driven by the pathogen burden.

Antigenic cartography of human and swine influenza A(H3N2) viruses

Derek J. Smith*†, Jan C. de Jong†, Alan S. Lapedes‡, Terry C. Jones*§, Colin A. Russell*, Theo M. Bestebroer†, Guus F. Rimmelzwaan†, Albert D. M. E. Osterhaus† and Ron A. M. Fouchier†

**Department of Zoology, University of Cambridge, Downing Street, Cambridge CB2 3EJ, UK, †National Influenza Centre and Department of Virology, Erasmus Medical Centre, Dr. Molewaterplein 50, 3015GE Rotterdam, NL, ‡Theoretical Division, T-13, MS B213, Los Alamos National Laboratory, Los Alamos, NM 87545, USA and §Departament de Tecnologia, Universidad de Pompeu Fabra, Passeig de Circumvallació 8, 08003 Barcelona, Spain*

Abstract. Influenza viruses are classic examples of antigenically variable pathogens and have a seemingly endless capacity to evolve in order to evade the immune response. The degree to which immunity induced by one strain is effective against another is mostly dependent on the antigenic difference between the strains. Antigenic drift is thus both the root cause of the enormous public health burden of influenza epidemics, and a primary reason why the virus is such a fascinating pathogen from a scientific perspective. Antigenic cartography is a new technique that can be used to analyse binding assay data and to obtain a detailed quantitative visualization of antigenic differences among pathogens. We provide a brief summary of antigenic cartography and its use to analyse the antigenic drift of influenza A(H3N2) viruses in humans and swine.

2008 Novel and re-emerging respiratory viral diseases. Wiley, Chichester (Novartis Foundation Symposium 290) p 32–44

Antigenic cartography is the process of applying mathematical, computational, and statistical techniques to antigenic binding assay data to create antigenic maps (Smith et al 2004a). These are not maps in the strict geographical sense, but in a biological sense: they provide a spatial layout of assay components (virus strains and antisera in the case of influenza), allowing the precise measurement of distances and directions among components. This gives a visualization of the underlying data and, more importantly, provides a concrete mathematical foundation for the quantitative analysis of antigenic data.

Antigenic differences between influenza viruses are routinely measured using the haemagglutination inhibition (HI) assay. The HI assay is a binding assay based on the ability of haemagglutinin (HA), a surface glycoprotein of the influenza virus, to agglutinate red blood cells, and the complementary ability of animal antisera raised against the same or related strains to block this agglutination (Hirst 1943). Thus an HI titre gives information about the affinity of an antiserum for a virus strain. One can interpret a titre value as rough measure of distance between antiserum and the virus. Tables of HI data, comprising titres of multiple antigens measured against multiple sera are notoriously difficult to interpret quantitatively due to paradoxes and irregularities in the data. Difficulties include some antisera being able to see differences between two antigens while other antisera cannot, and heterologous titres sometimes being higher than homologous titres. Previous methods used to analyse antigenic data quantitatively (Weijers et al 1985, Dekker et al 1995, Alexander et al 1997) have mostly been based on the methods of, or equivalent to, numerical taxonomy (Sneath & Sokal 1973). However, these methods sometimes give inconsistent results, improperly interpret data below the sensitivity threshold of the assay, and approximate antigenic distances in indirect ways. Lapedes & Farber (2001) introduced a geometric interpretation of HI data, that when combined with the methods of Smith et al (2004a) allowed the construction of antigenic maps from influenza HI data with a resolution of 0.83 (standard deviation 0.67) antigenic units (each unit corresponds to a twofold dilution of antiserum in the HI assay) and a correlation between predicted and measured values of 0.80 ($P \ll 0.01$).

Antigenic maps can be constructed in any number of dimensions and with any metric; for example, in a two dimensional plane, a 3D space, or spaces with higher dimensionality. Surprisingly, the HI data for human influenza subtype A(H3N2) allow accurate placement of strains and antisera into a 2D plane—the antigenic maps are just like the normal geographic maps that we are accustomed to reading, and so the HI data can be easily visualized. The visualizations provide a simple and intuitive way to assimilate large quantities of data.

Plate 1 is an antigenic map of 35 years of antigenic evolution of influenza A(H3N2) virus, from its emergence in humans in 1968 until 2003. This map was generated from a subset of the HI data measured from viral isolates collected as part of routine influenza surveillance in The Netherlands. Virus strains are shown as coloured shapes, and antisera as open shapes.

The map reveals both high-level and detailed features of the antigenic evolution of influenza A(H3N2) virus. The strains tend to cluster, like an archipelago of islands, rather than forming a continuous antigenic lineage, and the order of clusters, reading down the map, is mostly chronological, from the original Hong Kong 1968 cluster to the recent Fujian 2002 cluster (Smith et al 2004a, b). Each cluster is approximately the same size, the distances between clusters are fairly consistent, and the viral HA has drifted at approximately the same rate from 1968 to 2002.

Though the antigenic evolution of influenza A(H3N2) virus appears clustered, the genetic evolution of the virus appears gradual and continuous, suggesting that some amino acid substitutions have little antigenic effect (Smith et al 2004a). In analysis of the amino acid substitutions that define antigenic clusters it appears that a minority of amino acid substitutions contribute disproportionately to antigenic change.

Since 1982, the genetic evolution of swine and human H3N2 viruses, in terms of nucleotide and amino acid substitutions, have occurred at approximately the same rate (Smith et al 2004a, de Jong et al 2007). However, in sharp contrast, the rates of antigenic change of swine and human H3N2 viruses were very different. While human H3N2 viruses have evolved at a rate of about 2.0 antigenic units per year since 1982, swine H3N2 viruses have evolved more than six times as slowly, about 0.3 antigenic units per year (de Jong et al 2007). Thus, compared to human H3N2 viruses, swine H3N2 influenza viruses have accumulated many more nucleotide and amino acid substitutions in HA1 that had little effect on the antigenic properties of the virus (Plate 2).

Because of the short average life span of pigs, swine influenza virus evolution may be determined to only a limited extent by immune pressure, the driving force of antigenic drift of influenza viruses in humans. Whereas the human H3N2 virus requires frequent antigenic changes of HA to ensure that a sufficiently large pool of immunologically susceptible hosts is available, the vast majority of pigs are killed at the age of 6 months, so the susceptible pig population is continuously renewed, limiting the build-up of immune pressure. Only adult sows used for breeding live long enough to experience more than one influenza season, and thus, we speculate that these animals are essential for the endemicity of swine influenza viruses and may create some degree of immune pressure, leading to the (slow) antigenic drift.

Antigenic cartography is now routinely used in the human influenza vaccine strain selection process at the World Health Organization and has also been applied to avian and equine influenza viruses, rabies virus, and malaria parasites. There are no assumptions that limit the use of this methodology to the analysis of influenza virus and HI data, or infectious disease data—theoretically antigenic cartography could be used for any receptor–ligand system. Free and open source antigenic cartography software will be available from *http://www.antigenic-cartography.org*.

Acknowledgements

DJS was supported by the NIH Director's Pioneer Award Program, part of the NIH roadmap to medical research, through grant number DP1-OD000490-01. The research of ASL was supported by the Department of Energy under contract DE-AC52-06NA25396. The hospitality of the Santa Fe Institute where part of this work was performed is gratefully acknowledged.

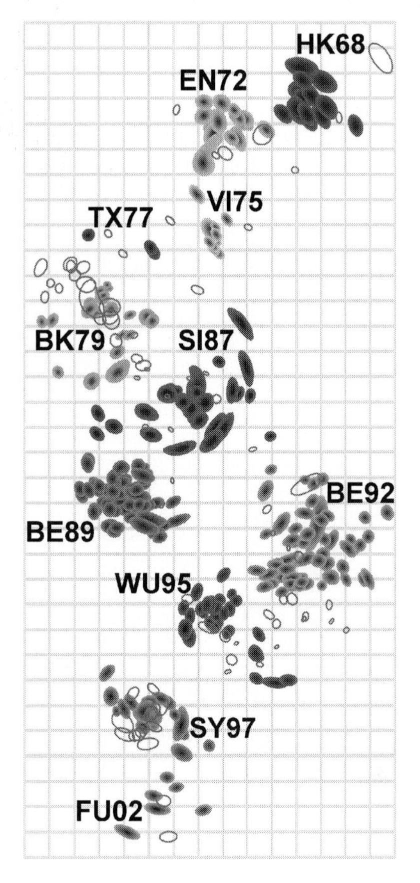

PLATE 1. An antigenic map of influenza A (H3N2) virus from 1968 to 2003. The relative positions of strains (coloured shapes) and antisera (uncoloured open shapes) were adjusted such that the distances between strains and antisera in the map represent the corresponding HI measurements with the least error. The periphery of each shape denotes a 0.5 unit increase in the total error; thus, size and shape represent a confidence area in the placement of the strain or antiserum. Clusters were identified by a k-means clustering algorithm and named after the first vaccine strain in the cluster—two letters refer to the location of isolation (Hong Kong, England, Victoria, Texas, Bangkok, Sichuan, Beijing, Wuhan, Sydney, Fujian) and the two digits refer to year of isolation. Strain colour represents the antigenic cluster to which each strain belongs. The vertical and horizontal axes both represent antigenic distance, and, because only the relative positions of antigens and antisera can be determined, the orientation of the map within these axes is free. The spacing between grid lines is 1 unit of antigenic distance—corresponding to a twofold dilution of antiserum in the HI assay. (Figure reprinted from Smith 2004a, with permission.)

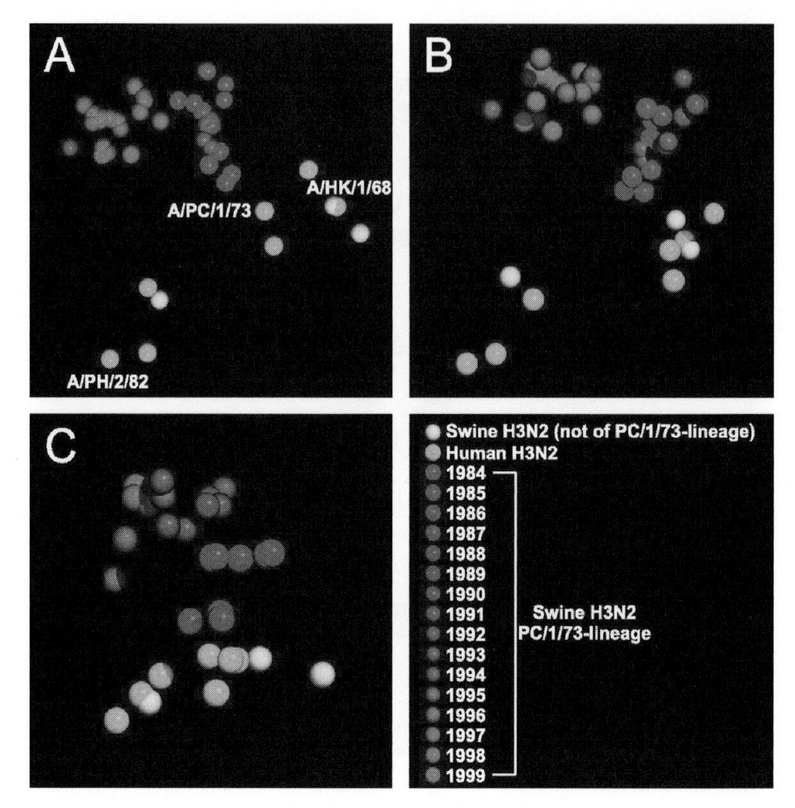

PLATE 2. Antigenic and genetic maps of swine and human H3N2 viruses. The relative positions of strains (coloured spheres) and antisera (not shown) in the maps were computed such that the distances between strains and antisera represent the corresponding HI measurements with the least error. Because only the relative positions of antigens can be determined, the orientation of the map within the axes is free. A two-dimensional map was generated on the basis of 20 HI tables, which included 30 ferret antisera and 48 influenza H3N2 viruses (A). The same analysis was repeated for the same strains to allow antigens and antisera to be positioned in a three-dimensional space (B). For the strains shown in panels A and B for which the HA1 amino acid sequence had been determined, a genetic map was generated based on an amino acid Hamming distance matrix (C). The colour-coding of the viruses is the same in the three maps, and is explained in the lower right panel. (Figure reprinted from de Jong 2007, with permission.)

References

Alexander DJ, Manvell RJ, Lowings JP, Frost KM, Collins MS 1997 Antigenic diversity and similarities detected in avian paramyxovirus type 1 (Newcastle disease virus) isolates using monoclonal antibodies. Avian Pathol 26:399–418

de Jong JC, Smith DJ, Lapedes AS, Donatelli I, Campitelli L 2007 Antigenic and genetic evolution of swine influenza A (H3N2) viruses in Europe. J Virol 81:4315–4322

Dekker A, Wensvoort G, Terpstra C 1995 Six antigenic groups within the genus pestivirus as identified by cross neutralization assays. Vet Microbiol 47:317–329

Hirst GK 1943 Studies of the antigenic differences among strains of influenza A by means of red cell agglutination. J Exp Med 78:407–423

Lapedes A, Farber R 2001 The geometry of shape space: application to influenza. J Theor Biol 212:57–69

Smith DJ, Lapedes AS, de Jong JC, Bestebroer TM, Rimmelzwaan GF 2004a Mapping the antigenic and genetic evolution of influenza virus. Science 305:371–376

Smith DJ, Lapedes A, de Jong JC, Bestebroer TM, Jones TC 2004b Mutations, drift, and the influenza archipelago. Discov Med 4:371–377

Sneath PHA, Sokal RR 1973 Numerical taxonomy. San Francisco: Freeman

Weijers TF, Osterhaus A, Beyer WEP, Vanasten J, Derondeverloop FM 1985 Analysis of antigenic relationships among influenza virus strains using a taxonomic cluster procedure—comparison of 3 kinds of antibody preparation. J Virol Methods 10:241–250

DISCUSSION

Webster: You have been successfully making predictions about human seasonal influenza for which strains of H1N1, H3N2 or B to use in influenza vaccines. There are approximately four distinct H5N1 viruses circulating in different parts of the world, can you predict which clade or subclade will successfully spread in humans?

Smith: We are not predicting where the evolution is going to go. In the prediction experiments I showed we were predicting HI titres, not predicting which new strains are going to emerge. If we'd never measured the titre between this serum and this antigen, we would be able to predict what the titre is from the antigenic map.

Osterhaus: What you are measuring is evolution in birds, not in humans. There is no selective drive in humans. Then it is important to look at the different bird populations for pre-existing immunity, again stressing how important it is to look in the bird population, not only at the viruses but also at their immunity level.

Smith: We don't know whether or not this evolution is being selected for by immunity. These antigenic changes may just be hitch-hiking along with other changes of the virus as it goes from host to host.

Osterhaus: Why has the virus become so rapidly evolving? Until 2002 nothing was happening. Is it because of spread?

Smith: There is a lot more spread. We do not know if that is the cause.

Skehel: You have much more H5 than you ever have had in circulation, so the possibility for subsequent infections is higher. There are more viruses around.

Osterhaus: If you do systematic surveillance, 10–20% of your birds may be positive for AI. These viruses may be replicating faster, but they are also dead ends. There has always been a lot of H5.

Skehel: Not at this level. You have been isolating H5 in all these countries Do you think it is just that surveillance has increased to show this?

Osterhaus: Low-path viruses are there all the time.

Kawaoka: Not at this magnitude.

Osterhaus: We see peaks now with high-path H5, but H5 and H7 have been there all the time.

Kawaoka: All the time, but not at this magnitude.

Osterhaus: The magnitude we see is only high-path; the low-path have always been under the radar screen in the wild birds. I am talking about wild birds only here.

Holmes: Do you have any sense for what the rules are for how the virus moves across the antigenic surface? What are the governing principles for this antigenic drift? Take human H3N2, do you have any notion it turns one way or another?

Smith: This is one of the things we are investigating. If the virus is evolving to keep away from prior immunity, then the best way to get away from it is to keep going in a straight line. Sometimes when we look at the evolution of H3, it will make a turn. There are some hints from those turns about what is going on. I pointed out that the Beijing 89 cluster was an evolutionary dead end. This cluster transition was caused by a single substitution, N to K at 145. But this was a dead end: the virus didn't evolve from that. However, it went on to another cluster, and then went off in the same direction again. It did N145K again but with three or four other substitutions.

Holmes: What does the NA do at the same time?

Smith: We don't know what the NA is doing.

Osterhaus: There is also T cell immunity and fitness. There are many other factors.

Smith: There are also potential issues with the stability of the haemagglutinin. Perhaps the virus can't continue to evolve from that place, and that particular natural experiment, going into a dead end and going through it again, is indicative of the fact that perhaps some of those co-mutations are required.

Skehel: In terms of your vaccine recommendations, have you looked at how they did in the past? There were some mistakes. Would you have made the same mistakes with the reagents available at that time?

Smith: Most of the vaccine missmatches had to do with viruses not being available at the time of vaccine strain selection. The Fujian update for example. There are very good explanations for the most missmatches.

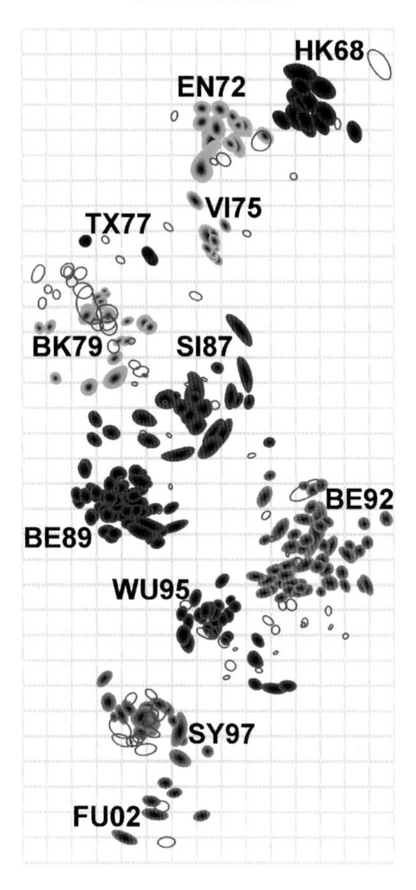

PLATE 1. An antigenic map of influenza A (H3N2) virus from 1968 to 2003. The relative positions of strains (coloured shapes) and antisera (uncoloured open shapes) were adjusted such that the distances between strains and antisera in the map represent the corresponding HI measurements with the least error. The periphery of each shape denotes a 0.5 unit increase in the total error; thus, size and shape represent a confidence area in the placement of the strain or antiserum. Clusters were identified by a k-means clustering algorithm and named after the first vaccine strain in the cluster—two letters refer to the location of isolation (Hong Kong, England, Victoria, Texas, Bangkok, Sichuan, Beijing, Wuhan, Sydney, Fujian) and the two digits refer to year of isolation. Strain colour represents the antigenic cluster to which each strain belongs. The vertical and horizontal axes both represent antigenic distance, and, because only the relative positions of antigens and antisera can be determined, the orientation of the map within these axes is free. The spacing between grid lines is 1 unit of antigenic distance—corresponding to a twofold dilution of antiserum in the HI assay. (Figure reprinted from Smith 2004a, with permission.)

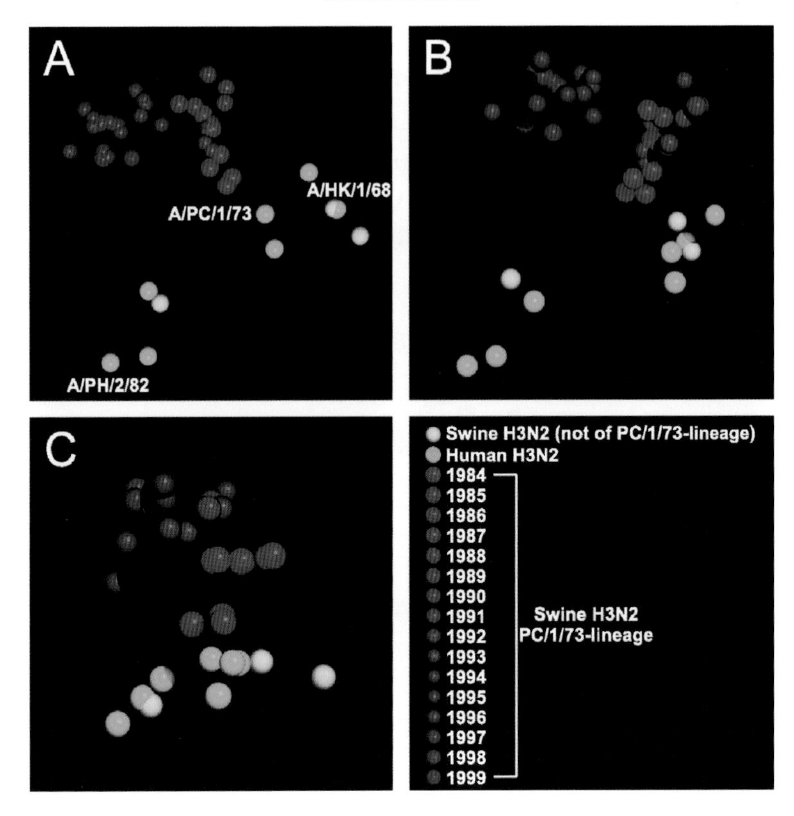

PLATE 2. Antigenic and genetic maps of swine and human H3N2 viruses. The relative positions of strains (coloured spheres) and antisera (not shown) in the maps were computed such that the distances between strains and antisera represent the corresponding HI measurements with the least error. Because only the relative positions of antigens can be determined, the orientation of the map within the axes is free. A two-dimensional map was generated on the basis of 20 HI tables, which included 30 ferret antisera and 48 influenza H3N2 viruses (A). The same analysis was repeated for the same strains to allow antigens and antisera to be positioned in a three-dimensional space (B). For the strains shown in panels A and B for which the HA1 amino acid sequence had been determined, a genetic map was generated based on an amino acid Hamming distance matrix (C). The colour-coding of the viruses is the same in the three maps, and is explained in the lower right panel. (Figure reprinted from de Jong 2007, with permission.)

Skehel: In fact, the vaccine recommendations were doing fine without your analysis.

Smith: Yes. Using antigenic cartography we have the opportunity to do finer-grain analyses, with higher accuracy. For example we can look at the evolution within an antigenic cluster, whereas previously the resolution of the assay has only been enough to tell the difference between clusters. This increased resolution provides the opportunity to figure out the evolutionary dynamics and move from tracking the evolution to potentially predicting it. We can also do things such as control for avidity differences algorithmically.

Skehel: Does the work that Stephen Fazekas (Fazekas de St Groth 1978) has done on the so-called absorption mutants still fit into your interpretation of these different titres?

Smith: I would say that it doesn't. If there really is just a simple geometric interpretation like this, it just says that in some cases the antigen that is used to infect a particular bird can end up in different positions in the map. We don't know the immunological basis for this.

Skehel: I thought Fazekas did know. His version on the binding story was that there was a mutation in the receptor binding site that was influencing receptor binding. In the competitive binding assays that are used, you could get those sorts of differences in higher titre or lower titre depending on the particular strain, just because it is a competitive assay.

Webster: He was wrong. That didn't work.

Smith: There are some other things that are a little bit like that. For instance, some sera turn out to be non-discriminatory sera. One might have said that these two middle sera here would be non-discriminatory. In many cases there might be situations like that going on, but what we would predict from this geometric interpretation is that this serum can't see the difference between these three clusters, but if there were a virus that was down at a lower point the sera would be able to see the difference.

Anderson: In your prediction model, how many sera do you need to get good predictions? Do you increase the reliability of your predictions with increasing number of sera, provided that they are appropriately spaced (i.e. distinct antibody responses)?

Smith: Yes. There is a direct analogy here with things like GPS navigation. You want to triangulate with things that are nicely spaced. For a two-dimensional antigenic map like this, all you need is three sera that are well spaced. We have taken large HI tables and 15 sera, made antigenics, and selected a much smaller number of sera that are well spaced and the antigenic maps remade with these are almost exactly the same. More sera are only better if they are better spaced.

Anderson: Does this allow a simpler way to define antigenic change in influenza virus?

Smith: This is the simpler way to define antigenic differences between isolates. By just looking at the HI data one can get led astray by what looks like paradoxes. If we look just at the raw HI data we can have a strain in one place that has the same titre to a serum as a strain somewhere else. One needs to shift from looking at the data in a tabular fashion to looking at it geometrically.

Wood: One of the problems in comparing HI data from different labs, particularly HI data looking at animal flu viruses, is that some labs use ferrets, others use chickens and others use rabbits. Can your technique make it a more level playing field so we can compare data using sera from different species?

Smith: It allows us to test whether or not the antigenic maps made with ferret data give a similar relative position to the antigens to antigenic maps made with hyperimmune rabbit data. My colleague, Ron Fouchier, has done exactly this with the H5 antigenic map. Maps are made with different sera and then we look to see how similar the positions of the antigens are in those maps. We find that for hyperimmune rabbits, ferrets and ducks we get almost exactly the same antigenic maps. We have also tested these H5 maps to see whether using horse erythrocytes or turkey erythrocytes makes a difference: we find the maps come out the same. However, in the situation with strongly adjuvanted vaccination, if we have a very broad response we find a twofold difference is going to be a much larger distance in the antigenic map. Thus, there are some sera where one needs to have a different scale.

Osterhaus: The whole issue of fitness is important. If we look at recent epidemics, we see that the H3N2 in humans becomes really wimpy. Has it exhausted its possibility to make all these escapes, or is it at the border of what the virus can do, and this is its limit

Smith: Presumably, there is some loss of intrinsic fitness when one makes a cluster transition. If there weren't, the virus would just evolve more quickly. The fact that it doesn't evolve more quickly is a major puzzle in terms of the emergence of new clusters.

Holmes: H1N1 has been around since 1977. Have you looked at the maps for this as well?

Smith: Not in detail. In terms of the epidemiology it is getting wimpier, but I don't know whether it is in terms of severity of disease. The mortality numbers seem to be coming down. But the rate of evolution hasn't changed over the years. The virus is potentially losing intrinsic fitness, but the gaps between clusters and the average amount of time between them is not changing. Surprisingly to us, it has not changed as a result of vaccination.

Webster: Are these flu viruses becoming wimpier? Is there a mechanism for becoming wimpy?

Kahn: There are certainly examples of pandemic flu viruses disappearing, such as H2. Does this model explain that at all?

Smith: The model I described is just representing antigenic differences among strains. If there were an assay for some other aspect of fitness, then one could make maps on wimpiness as well. In this particular assay we are looking at HI titres, so it isn't measuring fitness.

Anderson: John Wood, you raised the issue of the difficulty in comparing data from one sera to another sera. Derek Smith, your data suggest that if the sera are appropriately selected, then it shouldn't matter. Does this mean that the analysis and interpretation of the data is the issue?

Smith: HI tables from two different labs can look quite different.

Wood: That's why I made the point.

Anderson: It may be that this type of analysis would solve some the problems in comparing data between laboratories.

Smith: If one looks not in H5 where there are multiple species involved, but in terms of H3, we would predict that two tables that look very different in terms of titres would actually produce antigenic maps in which the relative positions of the antigens were the same, but the positions of the sera came out differently. Imagine the differences in the Dutch citiers example in my talk. If you chose a different set of reference points, the distances would all look different. But once you make the maps the positions of the cities would be the same but the reference points would differ.

Osterhaus: Even if you make mistakes they will be equalized because you have so many reference points.

Smith: Yes, that is also true within one map. The other test by fire of this method is that we process data from four WHO influenza collaborating centres. We ask how the data compares among the four centres, and we also compare these with the data collected in The Netherlands. The maps agree well.

Skehel: In the overlap and the selection of three sera, as you suggest, you could easily throw away information that might be apparant to an expert interpreting a group of HI tables.

Smith: If there is information in more sera than just three, the antigenic maps would come out differently. When there are more sera, although the titres look somewhat arbitrary, the data are all interconnected. They can be reduced in the case of H3 to something that is two dimensional.

Skehel: Do you get into debates with the people who are making the recommendations on the relative value of your maps versus their interpretation of HI data?

Smith: We agree shockingly well. John Wood and I both sit on this committee.

Wood: It seems as though the experts can make these maps in their heads!

Webster: At the last of those meetings there was an argument about H3N2, and whether they were going to become more Nepal-like, or stay with California. What is it doing?

Smith: Since the last strain selection meeting there has not been a jump to Nepal-like viruses, but to a different cluster.

Webster: Can you predict the future?

Smith: One of the main focuses of our research is to try to figure out just how predictable the evolution is. Is it predictable at all? It may not be. We now have one extra view on the evolution. For some years we have had the genetics, but we now have a couple of extra things that can be added to that. One is the quantification of antigenic relationships. The other is looking at the selection pressures on the virus with the sort of sera that have been collected by people like John Wood to try to see what the immune pressures are on the virus. If those pressures are the most important thing, it is possible that the evolution will be somewhat predictable. It is a little like predicting the weather. It is easier to predict short range than long range weather. We know there are aspects that are predictable. The great thing about influenza is that there have been a lot of data collected over many years, so one can take a data set up to a particular moment and then see how good prediction is from that point. There are excellent data going back 40 years for H3 which gives us the opportunity to test any predictive methods.

Holmes: There is one extra bit of complexity. You mentioned the genetics: what has been missing is the genomics. We have the whole genome story going back to 1992. Every single cluster jump from Beijing 1992 onwards has had a reassortment event co-occurring. This doesn't mean to say that this is the cause of it; they occur all the time. But what is happening is that the HA and NA genes are getting put in different genetic backgrounds all the time.

Osterhaus: Do you see them more frequently?

Holmes: I can't answer that. Again, I am not saying this is the cause. But the genome data are telling us that this virus reassorts continually. The HA is jumping around in different genetic backgrounds. This may explain the complicated pathways it is taking through space. If we add genome data, then we are a lot closer to making a predictive tool.

Peiris: I'd like to return to the issue of the 'wimpiness' of the virus. When we look at antigenic changes, it is progressively changing fairly evenly. But for the last 5 or 6 years those of us who are trying to do clinical trials on flu don't have enough flu in humans to do the trials on. Since the A/Sydney/5/07 drift variant emerged in 1997/8, there hasn't been a really major outbreak of flu. Have there been no reassortments over the last few years?

Holmes: The Fujian Californian jump was a genome-wide selective sweep. A whole new segment swept through the population at that time. Genetically that is a big change, and trying to translate it to fitness is complicated. This virus is reassorting all the time.

Peiris: How did the amantadine resistance mutation get fixed in human flu viruses so rapidly across the world?

Holmes: I have a theory: I think it was linked. There was a four by four segment reassortment event involving the amantadine resistant lineage. This mutation was probably fixed by hitchhiking with something else.

Osterhaus: What is the selective advantage there?

Holmes: That is a good question. This mutation has been fixed extremely quickly in countries that don't use amantadine often, so there is no selection process for it. It is a marker for something else. The genesis of that resistant lineage is by reassortment, which allowed it to get somewhere in evolutionary space it wasn't before, and something there is increasing the fitness.

Anderson: You could almost think of this line as an evolutionary blind alley. It is forced down an alley, and is accumulating antigenic change in a forced direction. This leaves it facing a brick wall where opportunities for lethal mutations are higher than advantageous mutations. Is that why H2N2 died out?

Holmes: I'm guessing that H2N2 died out because it was outcompeted. You are only low fitness if there is something of higher fitness competing with you. Fitness is dependent on competition. I think H1N1 and H2N2 are interesting: H1N1 has been around for a long time at low frequency. It is hanging around.

Skehel: The idea that fitness changes with receptor binding was put forward by Anne Underwood a long time ago (Whittaker & Underwood 1980). Her idea was that because the antigenic sites surround the receptor binding site, the introduction of substitutions in those regions compromised receptor binding increasingly with time during the pandemic. She presented reasonable HI data for this.

Osterhaus: Given that each year only 10% gets infected, how can the selective advantage be imagined? I would say it is T cell immunity.

Holmes: It can only be immune driven. The only trait that has continued to change since 1968 on a regular basis must be something that is immune driven.

Smith: When we look in pigs there is almost no antigenic evolution. There is no build up of immunity in the population.

Holmes: I can't believe any trait would change that quickly to select rapid evolution apart from immunity.

Lane: What about global air travel frequency as a way of creating a bigger global pool of immunity? Is it Ryanair that is saving us, making flu less virulent?

Smith: I think that's a great point.

Osterhaus: Flu goes all over the world anyway.

Su: Our government in Taiwan decided to produce its own influenza vaccine based on locally identified strains. This is on the basis of surveillance data which show that the strains circulating in Taiwan are always two years ahead of the WHO strains.

Smith: One of the great things about the global network is that we have data from around the world. We do see heterogeneity. We are actively looking at that spatial information.

Su: The data have been published (Shih et al 2005).

Tambyah: Has there been any impact of the use of antivirals on genetic or antigenic evolution? In Japan they are used very widely (Hatakeyama et al 2007).

Smith: We don't see any unusual patterns in the Japanese data. They seem to be tracking everywhere else in the world. I'd like to ask about the evolution of H5 and what might be driving that. Perhaps there is now the opportunity for birds to be reinfected, so there is a selective advantage for viruses that move. How much vaccine use is there, and might this be driving evolution?

Webster: It is antibody driven, but also this virus is moving between hosts at a grand rate. There are mammalian hosts and a huge number of avian hosts. Even within the same host it is changing. It is also reassorting like crazy.

Holmes: There is also segmental adaptation.

Webster: The chances are that the poor vaccines that are being used in poultry in some areas are driving it too, but I suspect it is a combination of factors.

Peiris: I think there's another issue. You mentioned that the early H5 viruses have not changed much. All the low path H5 viruses circulate in duck populations. Those high-path H5 outbreaks in poultry that have emerged in the industrialized world have been rapidly contained. There has never been such a prolonged Hi-path avian flu outbreak like this, one that has been going on for 10 years and which is so geographically widespread. This has allowed the virus the opportunity to diversify. You are seeing different populations of virus becoming dominant, and then you have more human cases in a geographic area. This may contribute to the diversity you are seeing. The fact that we have seen this unprecedented outbreak over a decade of a particular lineage of H5 has allowed an extent of diversity that has never happened before.

References

Fazekas de St Groth S 1978 Antigenic, adaptive and adsorptive variants of the influenza virus haemagglutinin. In: Laver WG, Bachmayer H, Weil R (eds) Topics in infectious diseases, Vol 3. Springer-Verlag, Wien, p 25–48

Hatakeyama S, Sugaya N, Ito M et al 2007 Emergence of influenza B viruses with reduced sensitivity to neuraminidase inhibitors. JAMA 297:1435–1442

Shih S-R, Chen G-W, Yang C-C et al 2005 Laboratory-based surveillance and molecular epidemiology of influenza virus in Taiwan. J Clin Microbiol 43:1651–1661

Whittaker RG, Underwood PA 1980 A mechanism for influenza subtype disappearance. Med Hypotheses 10:997–1008

Influenza pandemics and control

Gabriele Neumann* and Yoshihiro Kawaoka*†‡§

**Department of Pathobiological Sciences, School of Veterinary Medicine, University of Wisconsin-Madison, Madison, WI, USA, †Division of Virology, Department of Microbiology, and ‡International Research Center for Infectious Diseases, Institute of Medical Sciences, University of Tokyo, Tokyo, Japan and § Core Research for Evolutional Science and Technology (CREST), Japan Science and Technology Agency, Saitama, Japan*

Abstract. Influenza virus outbreaks, on a local or global scale, present a continuous risk to humans. This risk has increased with the finding that avian influenza viruses can transmit directly to humans and cause high mortality rates. Major efforts are, therefore, underway to better understand the pathogenicity of influenza viruses. Classic virology, reverse genetics, pathology studies and microarray approaches have revealed that aberrant immune-related gene responses may account for the unprecedented pathogenicity of the 'Spanish influenza' virus that killed an estimated 40 million people in 1918/1919. Influenza virus outbreaks can be controlled by vaccines and/or antiviral compounds. In the event of another pandemic, the efficient production of vaccine viruses will be critical. Current vaccine candidates to highly pathogenic avian H5N1 viruses do not grow efficiently in embryonated chicken eggs. Here, we present strategies for improving the growth of such vaccine candidates. We also discuss the recent emergence of resistance to neuraminidase inhibitors, including neuraminidase-resistant H5N1 viruses and highlight how, in a pandemic situation, current antivirals may provide only limited protection.

2008 Novel and re-emerging respiratory viral diseases. Wiley, Chichester (Novartis Foundation Symposium 290) p 45–55

Influenza viruses are one of the leading causes of respiratory disease in humans (for reviews, see Palese & Shaw 2007, Wright et al 2007). Typically, influenza epidemics in humans are caused by viral variants with point mutations in their haemagglutinin (HA) and neuraminidase (NA) proteins that allow the viruses to escape the immune response (antigenic drift); worldwide epidemics (pandemics) are caused by viruses with novel HA (or HA and NA) subtypes (antigenic shift).

The *in toto* introduction of avian influenza viruses into human populations was long considered a negligible risk; however, two major events exemplify the potential for just such an event. First, the 'Spanish influenza' of 1918/1919 was caused by an avian influenza virus that directly or indirectly transmitted to humans. Second, the emergence of highly pathogenic avian H5N1 influenza viruses, which first caused an outbreak in humans in Hong Kong in 1997. Over the past decade, these viruses have become enzootic in poultry populations in Southeast Asia, spread to Europe and Asia, and continue to infect humans with a mortality rate that exceeds 50%.

Two countermeasures exist to combat influenza virus outbreaks—vaccines and antivirals. Vaccines can be classified as either inactivated or live attenuated. Inactivated vaccines contain the HA and NA genes of the circulating wild-type viruses chosen for vaccine production, and all or some of their remaining genes from the A/Puerto Rico/8/34 (PR8, H1N1) virus. The PR8 virus genes allow efficient vaccine virus growth in embryonated chicken eggs. After chemical inactivation, the vaccine is administered as a split or subunit vaccine. Like inactivated vaccines, the live attenuated vaccines currently available in the US possess the HA and NA genes of the circulating wild-type viruses chosen for vaccine production; however their remaining genes come from the attenuated A/Ann Arbor/6/60 (H2N2) or B/Ann Arbor/1/66 viruses, whose internal genes provide the cold-adapted, temperature-sensitive, and attenuated phenotypes. Vaccine production typically takes 3–6 months after the selection of the vaccine strains. In the event of a pandemic, antivirals would therefore be the first line of defence. Two classes of antiviral compounds are available for human use—ion channel inhibitors, such as amantadine and rimantadine (which are effective against influenza A viruses only), and neuraminidase inhibitors such as Tamiflu® (oseltamivir) and Relenza® (zanamivir) (which are effective against influenza A and B viruses). In treated patients or cell culture experiments, influenza A viruses rapidly acquire resistance to ion channel inhibitors. Indeed, a large percentage of currently circulating human influenza A viruses (Bright et al 2006) and some H5N1 viruses are now resistant to amantadine and rimantadine (Cheung et al 2006). Hence, ion channel inhibitors are no longer a reliable weapon in the fight against human influenza. Recently, variants with increased resistance to neuraminidase inhibitors have also been reported (de Jong et al 2005, Kiso et al 2004, Le et al 2005).

Here, we summarize our efforts to better understand and control influenza viruses by deciphering the determinants of pathogenicity, developing improved vaccines, and studying the emergence of variants with increased resistance to neuraminidase inhibitors.

The 'Spanish influenza' virus—decoding a deadly secret

The 'Spanish influenza' is the deadliest infectious disease recorded in human history and reduced the life expectancy in the USA by more than 10 years. The causative virus was only recently made available when Taubenberger and colleagues isolated pieces of viral RNA from tissue samples and corpses of victims of the 'Spanish influenza' (Basler et al 2001, Reid et al 1999, 2000, 2002, 2004, Taubenberger et al 2005). Amplification of these fragments by RT-PCR yielded the genetic information of all eight segments. Based on this information, the 'Spanish influenza' virus was recreated by reverse genetics (Tumpey et al 2005), a technology for the generation of influenza viruses from cDNA (Neumann et al 1999). The reconstituted virus was highly pathogenic in mice and embryonated chicken eggs and grew to high titres in human bronchial epithelial cells (Tumpey et al 2005). Microarray studies of infected mice showed increased activation of the genes that regulate the host immune response (Kash et al 2006).

To better understand the pathogenicity of the 'Spanish influenza' virus, we also generated 'Spanish influenza' virus and tested it in cynomolgus macaques (Kobasa et al 2007). A contemporary human virus of the same subtype (A/Kawasaki/173/01; H1N1) served as a control. Infection of nonhuman primates with the reconstituted 1918 pandemic virus or the contemporary human virus caused macroscopic pathologic changes that were limited to the lungs. A prominent characteristic of lungs infected with 'Spanish influenza' virus was antigen in plump alveolar cells. These cells desquamated into the alveolar space. By day 6–8 postinfection, the lungs of animals infected with the contemporary human virus began to heal, evidenced by thickening of the alveolar wall and the absence of viral antigen. By contrast, in animals infected with the pandemic virus, severe lesions were observed by day 6–8 postinfection. There was also extensive oedema and haemorrhagic exudate, as was reported for patients who succumbed to the 'Spanish influenza'. These pathological findings correlated with microarray studies that revealed significant differences between animals infected with the pandemic versus the contemporary human strain. In particular, the 'Spanish influenza' virus caused sustained high levels of immune-related genes, such as cytokines and chemokines, which were not seen with the contemporary human virus (Kobasa et al 2007), suggesting modulation of the host immune response. Hence, an aberrant immune response may explain, at least in part, the severity of the 'Spanish influenza' virus infection. Future studies will include a more detailed analysis of the host immune response, and the identification of the viral determinants of the pathogenicity of the 'Spanish influenza' virus.

Improved H5N1 vaccine viruses

Highly pathogenic avian H5N1 influenza viruses continue to infect humans, causing worldwide concerns of a pandemic and accelerating the development of H5N1 vaccines. These viruses kill embryonated chickens, resulting in low virus yields. Moreover, they pose a risk for vaccine production staff. Reverse genetics has allowed the 'detoxification' of these highly pathogenic H5N1 viruses by replacing the multiple basic amino acids at the HA cleavage site (a major determinant of highly pathogenic influenza viruses) with an avirulent-type cleavage site. Several candidate vaccines have now been generated (Horimoto et al 2006, Nicolson et al 2005, Subbarao et al 2003, Webby et al 2004, Wood & Robertson 2004) that contain the NA and 'detoxified' HA genes of an H5N1 virus in the genetic background of the PR8 virus, the WHO-recommended 'backbone' strain. Some of these candidate vaccines, however, do not grow efficiently in embryonated chicken eggs. Combined with the limited immunogenicity of some current H5N1 vaccine candidates (Treanor et al 2006), substantial production efforts would therefore be needed in a pandemic situation to ensure sufficient amounts of viral antigen. One possible solution to this challenge is the development of vaccine candidates with more robust growth properties.

Since its initial isolation in 1934, the PR8 virus has been maintained in laboratories around the world. Accordingly, several variants now exist, each with their own growth properties. We compared the growth properties of the NIBRG-14 vaccine candidate (Nicolson et al 2005) [which possesses the A/Vietnam/1194/04 (VN1994) HA and NA genes in the background of the PR8(Cambridge) strain (George Brownlee, personal communication)] with a virus that possesses the same HA and NA genes as NIBRG-14 but its remaining genes derived from a PR8 strain maintained in our laboratory [designated PR8(UW)] (Horimoto et al 2007). We found that our vaccine candidate grew appreciably better in embryonated chicken eggs than did the original NIBRG-14 virus, suggesting that our PR8(UW) variant may be better suited for cost-efficient, large-scale vaccine production than PR8(Cambridge).

To identify the gene segment(s) that accounts for the superior growth properties of our vaccine candidate relative to NIBRG-14, we tested a number of reassortant viruses (Horimoto et al 2007), and found that the polymerase and nucleoprotein (NP) genes are critical for efficient growth in embryonated chicken eggs. Sequence analysis revealed a number of amino acid substitutions in the following proteins of our vaccine candidate—6 in PB2, 5 in PB1, 2 in PB1-F2, 2 in PA and 3 in NP.

The functional balance between the HA and NA proteins also likely affects the growth properties of vaccine viruses. We therefore generated a number of NIBRG14-UW variants in which the VN1194 NA gene was replaced with that of the following N1 strains: A/Vietnam/1203/04 (VN1203; H5N1), A/Hong Kong/213/03 (HK213; H5N1), A/Hong Kong/486/97 (HK486; H5N1), A/WSN/33 (WSN; H1N1), A/Kanagawa/173/01 (Kanagawa; H1N1), and PR8(UW). Of these, the variant possessing the PR8(UW) NA gene grew to significantly higher titres than NIBRG-14UW (Horimoto et al 2007). Similarly, we tested the above-listed NA genes in combination with the VN1203 HA gene in the genetic background of PR8(UW) and found that the variant possessing the PR8(UW) NA gene grew most efficiently (Horimoto et al 2007).

To assess if higher virus titres in eggs correlated with increased amounts of HA antigen, we determined the HA concentrations of all of the test viruses and found significantly higher amounts of HA for our vaccine candidate as compared to NIBRG-14 (Horimoto et al 2007).

We also found that a 7:1 reassortant [possessing only an H5 HA gene in combination with seven gene segments derived from PR8(UW)] grew significantly better in eggs than did the respective 6:2 reassortant. One might argue that replacing the H5N1 virus NA gene with that of PR8(UV) affects the immune responses to this virus. The major protective antigen in inactivated influenza vaccines, however, is HA, and a slightly reduced immune response may be offset by the ability to generate significantly higher amounts of vaccine virus. Such a consideration would be especially relevant to pandemic situations when embryonated chicken eggs will most likely be in short supply and efficient virus growth may be critical for rapid, large-scale vaccine production and distribution.

Resistance to neuraminidase inhibitors

In contrast to the rapid emergence of resistance to ion channel inhibitors, resistance to neuraminidase inhibitors requires multiple passages in cell culture in the presence of the compounds (Kimm-Breschkin 2000). Controlled clinical trials have demonstrated oseltamivir resistance in about 4% of children aged 1–12, and in 0.4–1% of adults (Hayden 2001, Treanor et al 2000, Whitley et al 2001).

To gain further insight into the frequency of oseltamivir resistance, we studied samples from 50 Japanese children (aged 2 months–16 years; median 3.7 years) who were treated with oseltamivir (Kiso et al 2004). For all patients, treatment began within 48 h of symptom onset. For each patient, pre- and post-treatment samples were available. Influenza virus was detected by RT-PCR in the post-treatment samples of 33 patients, but the infection had been cleared in the remaining individuals. The NA and HA genes of all samples were amplified by RT-PCR, cloned, and sequenced. Sequence comparison with pre-treatment samples revealed neuraminidase mutations in viruses isolated from 9 of the 50 oseltamivir-treated patients (18%) (Kiso et al 2004). Eight of these isolates carried mutations known to confer resistance to neuraminidase inhibitors (Gubareva et al 2000): an Arg292Lys mutation, which was found in six patients, and a Glu119Val mutation, found in two patients. One isolate possessed a mutation (Asn294Ser) that had not previously been associated with oseltamivir resistance.

To assess if the observed NA mutants did indeed confer resistance to oseltamivir, we carried out *in vitro* studies that determined the NA sensitivity to GS4071 (i.e., oseltamivir carboxylate) (Kiso et al 2004). All three mutations conferred NA resistance to GS4071: we detected a 10^4–10^5-fold increase for Arg292Lys, a 500-fold increase for Glu119Val, and a 300-fold increase for Asn294Ser. These findings confirm that the Arg292Lys and Glu199Val mutations confer resistance to oseltamivir. More importantly, they identified the Asn294Ser mutation as a previously unrecognized mutation that confers oseltamivir resistance.

Our study also demonstrated that oseltamivir-resistant viruses emerge more frequently than previously reported (Gubareva et al 2001, Hayden 2001, Zambon & Hayden 2001). The results of our study with young children who were likely encountering their first episode of influenza may mimic what would happen in a pandemic in which the population is immunologically naïve to the pandemic strain. Concerns over oseltamivir-resistance have further heightened with the isolation by us (Le et al 2005) and others (de Jong et al 2005, *http://www.who.int/csr/don/2007_01_18/en/index.html*) of oseltamivir-resistant H5N1 viruses from H5N1 virus-infected individuals who had been treated with the drug.

Neuraminidase inhibitors are also effective against influenza B viruses, but little is known about the frequency of influenza B virus resistance to neuraminidase inhibitors. In fact, only two influenza B viruses with reduced sensitivity to neuraminidase inhibitors have been described to date, both of which were isolated from immunocompromised children (Gubareva 2004, Gubareva et al 1998). An influenza B virus outbreak in Japan in the winter of 2004–2005 allowed us to study in more detail the frequency of emergence of variants with reduced sensitivity to neuraminidase inhibitors. For 65 sets of pre- and post-treatment samples obtained from children, we compared the *in vitro* sensitivity to oseltamivir and found one case (1.4%) in which the IC_{50} value of the post-treatment isolate was significantly higher than that of the pre-treatment sample (Hatakeyama et al 2007). Sequence analysis revealed a Gly402Ser mutation located near the active site of the NA protein.

We also determined the IC_{50} values of 422 samples obtained from untreated patients (Hatakeyama et al 2007). We identified 7 isolates with increased IC_{50} values. Sequence analysis of the NA proteins of these isolates revealed mutations in all 7 (3 with an Asp198Asn mutation, 3 with an Ile222Thr mutation, and 1 with a Ser250Gly mutation). It is likely that these cases reflect person-to-person transmission of variants with reduced sensitivity to neuraminidase inhibitors. It is important to note that variants with increased resistance to neuraminidase inhibitors have only appeared since the release of these drugs to the public, arguing against the spontaneous emergence in nature of oseltamivir- or zanamivir-resistant variants.

Collectively, our findings indicate that influenza B viruses with reduced sensitivity to neuraminidase inhibitors can emerge during oseltamivir or zanamivir treatment. The rate with which such variants arise is lower for influenza B viruses (1.4%) than for influenza A viruses (up to 18%; Kiso et al 2004). Moreover, our study suggests that influenza B variants with increased sensitivity can be transmitted among household members, and probably also within communities.

Summary

To combat influenza viruses, we need to better understand the mechanisms of pathogenicity, and to develop more efficient and efficacious vaccines and antiviral treatments. Our current studies with the 'Spanish influenza' virus and highly pathogenic H5N1 avian influenza viruses are expected to elucidate the viral and cellular determinants of pathogenicity. Such studies may identify novel targets for antivirals that are urgently needed as resistance to ion channel and neuraminidase inhibitors continues to emerge. Moreover, efforts are underway to increase the efficiency of vaccine virus production—an issue that will be of the utmost importance in the event of a pandemic.

Acknowledgements

We thank Susan Watson for editing the manuscript. This work was supported, in part, by grants-in-aid from the Ministries of Education, Culture, Sports, Science, and Technology, of Health, Labor, and Welfare of Japan, by CREST (Japan Science and Technology Agency), by a contract research fund from the Ministry of Education, Culture, Sports, Science and Technology, Japan, for Program of Founding Research Centers for Emerging and Reemerging Infectious Diseases, and by National Institute of Allergy and Infectious Diseases Public Health Service research grants, USA.

References

Basler CF, Reid AH, Dybing JK et al 2001 Sequence of the 1918 pandemic influenza virus nonstructural gene (NS) segment and characterization of recombinant viruses bearing the 1918 NS genes. Proc Natl Acad Sci USA 98:2746–2751

Bright RA, Shay DK, Shu B et al 2006 Adamantane resistance among influenza A viruses isolated early during the 2005–2006 influenza season in the United States. JAMA 295:891–894

Cheung CL, Rayner JM, Smith GJ et al 2006 Distribution of amantadine-resistant H5N1 avian influenza variants in Asia. J Infect Dis 193:1626–1629

de Jong MD, Tran TT, Truong HK et al 2005 Oseltamivir resistance during treatment of influenza A (H5N1) infection. N Engl J Med 353:2667–2672

Gubareva LV 2004 Molecular mechanisms of influenza virus resistance to neuraminidase inhibitors. Virus Res 103:199–203

Gubareva LV, Matrosovich MN, Brenner MK et al 1998 Evidence for zanamivir resistance in an immunocompromised child infected with influenza B virus. J Infect Dis 178:1257–1262

Gubareva LV, Kaiser L, Hayden FG 2000 Influenza virus neuraminidase inhibitors. Lancet 355:827–835

Gubareva LV, Kaiser L, Matrosovich MN et al 2001 Selection of influenza virus mutants in experimentally infected volunteers treated with oseltamivir. J Infect Dis 183:523–531

Hatakeyama S, Sugaya N, Ito M et al 2007 Emergence of influenza B viruses with reduced sensitivity to neuraminidase inhibitors. JAMA 297:1435–1442

Hayden FG 2001 Perspectives on antiviral use during pandemic influenza. Philos Trans R Soc Lond B Biol Sci 356:1877–1884

Horimoto T, Takada A, Fujii K et al 2006 The development and characterization of H5 influenza virus vaccines derived from a 2003 human isolate. Vaccine 24:3669–3676

Horimoto T, Murakami S, Muramoto Y et al 2007 Enhanced growth of seed viruses for H5N1 influenza vaccines. Virology 366:23–27

Kash JC, Tumpey TM, Proll SC et al 2006 Genomic analysis of increased host immune and cell death responses induced by 1918 influenza virus. Nature 443:578–581

Kimm-Breschkin JL 2000 Resistance of influenza viruses to neuraminidase inhibitors—a review. Antiviral Res 47:1–17

Kiso M, Mitamura K, Sakai-Tagawa Y et al 2004 Resistant influenza A viruses in children treated with oseltamivir: descriptive study. Lancet 364:759–765

Kobasa D, Jones SM, Shinya K et al 2007 Aberrant innate immune response in lethal infection of macaques with the 1918 influenza virus. Nature 445:319–323

Le QM, Kiso M, Someya K et al 2005 Avian flu: isolation of drug-resistant H5N1 virus. Nature 437:1108

Neumann G, Watanabe T, Ito H et al 1999 Generation of influenza A viruses entirely from cloned cDNAs. Proc Natl Acad Sci USA 96:9345–9450

Nicolson C, Major D, Wood JM et al 2005 Generation of influenza vaccine viruses on Vero cells by reverse genetics: an H5N1 candidate vaccine strain produced under a quality system. Vaccine 23:2943–2952

Palese P, Shaw M 2007 Orthomyxoviridae: the viruses and their replication. In: Knipe DM, Howley PM, Griffin DE et al (eds) Fields Virology. Wolters Kluwer, Lippincott Williams & Wilkins, Philadelphia, p 1647–1689

Reid AH, Fanning TG, Hultin JV et al 1999 Origin and evolution of the 1918 'Spanish' influenza virus hemagglutinin gene. Proc Natl Acad Sci USA 96:1651–1656

Reid AH, Fanning TG, Janczewski TA et al 2000 Characterization of the 1918 'Spanish' influenza virus neuraminidase gene. Proc Natl Acad Sci USA 97:6785–6790

Reid AH, Fanning TG, Janczewski TA et al 2002 Characterization of the 1918 'Spanish' influenza virus matrix gene segment. J Virol 76:10717–10723

Reid AH, Fanning TG, Janczewski TA et al 2004 Novel origin of the 1918 pandemic influenza virus nucleoprotein gene. J Virol 78:12462–12470

Subbarao K, Chen H, Swayne D et al 2003 Evaluation of a genetically modified reassortant H5N1 influenza A virus vaccine candidate generated by plasmid-based reverse genetics. Virology 305:192–200

Taubenberger JK, Reid AH, Lourens RM et al 2005 Characterization of the 1918 influenza virus polymerase genes. Nature 437:889–893

Treanor JJ, Hayden FG, Vrooman PS et al 2000 Efficacy and safety of the oral neuraminidase inhibitor oseltamivir in treating acute influenza: a randomized controlled trial. US Oral Neuraminidase Study Group. JAMA 283:1016–1024

Treanor JJ, Campbell JD, Zangwill KM et al 2006 Safety and immunogenicity of an inactivated subvirion influenza A (H5N1) vaccine. N Engl J Med 354:1343–1351

Tumpey TM, Basler CF, Aguilar PV et al 2005 Characterization of the reconstructed 1918 Spanish influenza pandemic virus. Science 310:77–80

Webby RJ, Perez DR, Coleman JS et al 2004 Responsiveness to a pandemic alert: use of reverse genetics for rapid development of influenza vaccines. Lancet 363:1099–1103

Whitley RJ, Hayden FG, Reisinger KS et al 2001 Oral oseltamivir treatment of influenza in children. Pediatr Infect Dis J 20:127–133

Wood JM, Robertson JS 2004 From lethal virus to life-saving vaccine: developing inactivated vaccines for pandemic influenza. Nat Rev Microbiol 2:842–847

Wright PF, Neumann G, Kawaoka Y 2007 Orthomyxoviruses. In: Knipe DM, Howley PM, Griffin DE et al (eds) Fields Virology. Wolters Kluwer, Lippincott Willams & Wilkins, Philadelphia, p 1691–1740.

Zambon M, Hayden FG 2001 Position statement: global neuraminidase inhibitor susceptibility network. Antiviral Res 49:147–156

DISCUSSION

Webster: The current way of making influenza vaccines is to make high growth reassortants by old fashioned mixed infection and selection. For H5N1 vaccines reverse genetics is used for vaccine preparation. Industry has decided that reverse genetics makes poor yielding vaccine strains. You show us clearly that you are defining the genes required for high growth. Is there high protein yield as well as HA activity?

Kawaoka: We examined the HA content of our vaccine strains and found that they do indeed have higher amounts of the HA protein.

Webster: It is important to convince industry. The future of flu vaccines lies with reverse genetics, in my opinion, both for seasonal and pandemic vaccines. This is really important.

Wood: The industry is seeing data largely from RG14, which is the virus they have mainly been using. Both the wild-type virus, A/Vietnam/1194/2004, and the RG14 virus have an unusual haemagglutinin content. It is very low compared with normal flu viruses.

Kawaoka: I understand that vaccine candidates with more recent H5N1 strains grow better.

Wood: Yes, they are more normal. This particular virus may just be an oddball. It is a shame that manufacturing industry has a poor view of reverse genetics just on the basis of one virus.

Webster: This is important information to get out there.

Osterhaus: I know you don't believe in T cell immunity, but given that all these RG viruses are largely based on the 6,2 RG systems, whereas we know that major T cell epitopes are most probably in the nuclear protein and possibly the matrix protein, wouldn't there be a case for looking into other possibilities and go for a 5,3 to at least include the nuclear protein.

Kawaoka: Yes, if we were developing live vaccines.

Osterhaus: Yes, but for CD4 responses it is a similar story.

Skehel: HA is a perfectly good immunogen for T cells that function as helper cells for antibody producing cells.

Osterhaus: As we said before, neutralizing antibodies may not be generated and still you have protection.

Kawaoka: I am not sure of the extent of NP's contributions to protective immunity in inactivated vaccines.

Osterhaus: If it is a PR8 background, it could be that the NP does not carry the CD4 epitopes that you might need from the avian virus strains.

Lai: In some patients you show there are drug-resistant mutants before treatment. Do they have the same mutation sites?

Kawaoka: Some viruses isolated before treatment had amino acid residues that are known to confer drug resistance.

Lai: Was this amplified after the treatment?

Kawaoka: No, it was the same. It was a pure population from the beginning. We cloned the NA gene, put it into a plasmid and calculated the frequency. Those patients who shed resistant viruses before treatment continued to shed 100% resistant viruses after treatment.

Peiris: In your 1918 studies you looked at microarrays in bronchus. Have you looked at defined cell populations to complement these data? When we look at animals we are looking at complex interactions.

Kawaoka: No. That would be the next step.

Osterhaus: The problem of isolating those cells is that they are no longer in the same native state.

Smith: Where did the zanamivir-resistant strain come from? What were the selection pressures for it?

Kawaoka: I believe that zanamivir-resistant viruses were selected during treatment in patients.

Smith: Was there any difference in the severity of the disease in those individuals who had resistant virus?

Kawaoka: It was the same.

Holmes: What is their fitness in the absence of the drug?

Kawaoka: For influenza A viruses, many of the mutations that confer NA-inhibitor resistance make the virus less fit, except the mutation at 119. For influenza B viruses, since they were isolated from patients who were not treated, they must be fit

Skehel: Is there any indication of mutations in HA that might give resistance?

Kawaoka: Most of these B viruses did not have mutations in their HAs, with one exception.

Webster: Is there any virus in the faeces of the macaques?

Kawaoka: We didn't find any.

Webster: In studies on ferrets, we get quite a lot of shedding of H5N1.

Osterhaus: H5 in monkeys doesn't end up in the faeces. Unlike the ferrets, there isn't the more generalized disease. 1918 looks similar to what we see in H5.

Webster: We need to be a little careful with the statement about H5N1 in the ferret. It is only a few strains—the highly pathogenic ones that cause systemic infection. I have a question about the use of statins for 1918. This is a question I always get asked: should we use them? What is your response?

Kawaoka: I heard that someone tested statins in mice, but it did not work.

Osterhaus: There may be recent unpublished data that look a little bit more favourable.

Kawaoka: I'm not convinced yet.

Kahn: What is the biological basis for why statins might work?

Hoffmann: It's because of their effect on the so-called 'cytokine storm'.

Peiris: Has anyone put the 1918 into chickens? Is it highly pathogenic for chickens? In the original paper (Tumpey et al 2005) the virus was reported to be trypsin-independent and lethal for chick embryos.

Skehel: 1918 is trypsin-dependent?

Kawaoka: If you use MDCK cells that don't produce proteases, they are trypsin-dependent. This makes sense since the HA does not have a series of basic residues at the cleavage site. I don't think it is pathogenic in chickens.

Reference

Tumpey TM, Basler CF, Aguilar PV et al 2005 Characterization of the reconstructed 1918 Spanish influenza pandemic virus. Science 310:77–80

On the activation of membrane fusion by influenza haemagglutinin

J. J. Skehel, S. Wharton, L. Calder and D. Stevens

Division of Virology, MRC National Institute for Medical Research, Mill Hill, London, UK

Abstract. We present data on haemagglutinin (HA) stability in relation to membrane fusion activation and activity. Membrane fusion activation occurs in two stages. Firstly, HA is primed for activation by cleavage of the primary product of HA mRNA translation, precursor HA0. Cleavage, which is essential for infectivity, occurs at arginine 329 or the equivalent residue in HA0 of fourteen of the sixteen antigenic subtypes of avian and human viruses, or at a furin recognition sequence, Arg-X-Arg/Lys-Arg, inserted at 329 in HA0 of highly pathogenic avian viruses of H5 and H7 subtypes. Secondly, the membrane fusion potential of cleaved HA is activated during infection in response to low pH, between pH 5 and pH6.5 depending on the strain of virus, in endosomes. As a consequence, the virus membrane is fused with the endosomal membrane resulting in the transfer of the genome-transcriptase core of the virus into the cell, and the initiation of virus replication. We have done experiments to compare the properties of uncleaved HA0 and cleaved HA with the objective of understanding more about the process of activation of membrane fusion. The results are in two sections concerning HA and HA0, respectively.

2008 Novel and re-emerging respiratory viral diseases. Wiley, Chichester (Novartis Foundation Symposium 290) p 56–68

Cleaved haemagglutinin (HA)

Incubation of virus or purified HA at fusion pH results in changes in the conformation of HA that are required for membrane fusion (Skehel & Wiley 2000). The nature of these changes has been analysed biochemically (Skehel et al 1982), antigenically (Daniels et al 1983), by electron microscopy (Ruigrok et al 1986, 1988), and by X-ray crystallography of protease fragments of HA (Bullough et al 1994, Bizebard et al 1995) (Fig. 1).

In addition to low pH it was observed that heating virus at neutral pH also activated membrane fusion (Wharton et al 1986, Ruigrok et al 1986, 1988) and that there was co-variation between the temperature and the pH at which fusion by mutant HAs was activated (Table 1). In similar experiments, incubation in urea at neutral pH (Table 1 and Carr et al 1997) also activated fusion by the mutants and again, the lower the pH of activation, the higher the temperature of activation and the higher the concentration of urea required for activation. The following

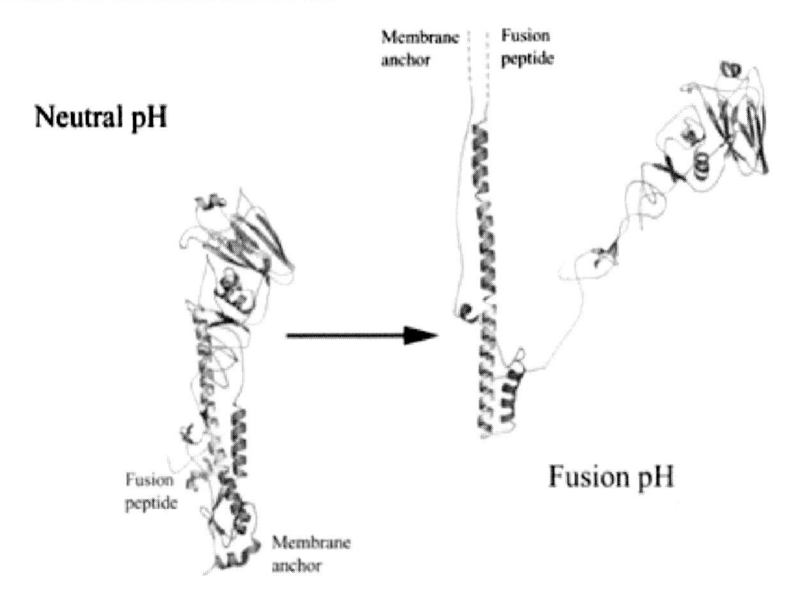

FIG. 1. Changes in the conformation of haemagglutinin required for membrane fusion activity. The HA monomer is colour-coded (not visible in this reproduction) to indicate the changes in location of different segments of structure on incubation at fusion pH. The data on which the diagram is based are from Wilson et al 1981, Bizebard et al 1995 and Bullough et al 1994.

TABLE 1 Co-variation of conditions for conformational changes in mutants of X-31 virus (H_3 subtype)

Virus	*pH*	*Heat (°C)*	*Urea (conc M)*
wt X-31	5.6	64	6.1
$D112_2G$	6.0	54	2.9
$H17_1R$	6.2	48	1.8

The pH of conformational change was determined using a virus-liposome fusion assay (Wharton et al 1986). The melting temperature is the mid-point temperature of the decrease in tryptophan fluorescence brought about by thermal unfolding. The urea concentrations are the mid-points of the decrease in tryptophan fluorescence after incubation in increasing concentrations of urea.

experiments were done to characterize the changes in HA structure that accompanied fusion activation.

(1) Digestion with trypsin following incubation at fusion pH generated two major fragments of HA1 of molecular weight 40 000 and 25 000 (Skehel et al 1982) (Fig. 2), and the same fragments were formed by proteolysis following incubation in urea. Both fragments were generated by cleavage at HA1 Lys27, the 40 000 Da fragment terminated at the C-terminus of HA1 and the 25 000 Da fragment terminated at Arg224 of HA1. Heating followed by proteolysis led to the formation

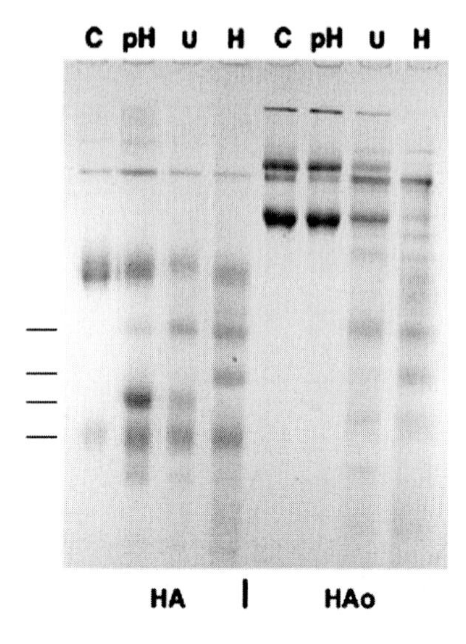

FIG. 2. Trypsin digestion of HA0 and HA following incubation at fusion pH, in urea, or heating. HA was incubated at pH 5, or in 8 M urea at pH 7, or heated at 70 °C. Samples were neutralized, urea removed by dialysis, and then digested with trypsin (1%) at 37 °C for 10 minutes. HA0 (the R329Q mutant used to prevent cleavage during preparation) was incubated at pH 5, or in 5 M urea at pH7, or heated at 60 °C. Samples were neutralized, urea removed by dialysis, and then digested with trypsin (0.5%) at 37 °C for 10 minutes. The fragments marked have approximate molecular weights of 40 000, 30 000, 25 000 and 20 000. C, control untreated samples; pH, samples incubated at pH 5; U, samples incubated in urea; H, heated samples.

TABLE 2 Comparisons of conditions for conformational changes in HA and HA_0 of X-31 virus (H_3 subtype)

Protein	*pH*	*Heat (°C)*	*Urea (conc M)*
HA	5.6	64	6.1
HA_0	no changes	52	2.4

Antibody binding was measured using a solid phase ELISA.

of fragments with molecular weights of 40 000 and 30 000, also generated by cleavage at HA1 Lys27.

(2) The antigenic properties of HA changed on activation of fusion as judged by reactivities with specific monoclonal antibodies. Antibodies that recognize HA1 residue 144 in a prominent loop near the membrane-distal tip of HA, bind indistinguishably, irrespective of pH or heating or incubation in urea. By contrast, antibodies that recognize the region that includes HA2 residue 107 and that forms a turn by re-folding of the central α-helix on activation of fusion, bind only to activated HA (Table 2).

(3) Electron microscopy studies have indicated that at fusion pH, HA rosettes change in structure having longer and thinner spike-like projections from bigger central aggregates than neutral pH rosettes (Ruigrok et al 1988). Similar structures are formed by incubation in urea (Fig. 3). The HAs in heated rosettes by contrast, appear swollen and less spike-like, as in the proteolysis experiments, suggesting a difference between the structural changes on heating from those at fusion pH and those on incubation in urea.

Uncleaved precursor haemagglutinin, HA0

The conformation of HA0 is unchanged on incubation at low pH from that at neutral pH. However, on heating or incubation in urea, as judged by any of the criteria used in the analyses of HA, changes in structure were detected. The changes occurred at lower temperature (52 degrees for HA0 vs. 64 degrees for HA), and at lower urea concentration (2.4 M for HA0 vs. 6.1 M for HA) (Table 3). Antigenically, heat or urea treatments resulted in reactivity with antibody 107, which recognizes the epitope formed in HA at fusion pH, suggesting that the changes in structure are similar. This is also the conclusion from proteolysis experiments (Fig. 2), since a prominent 40 000 molecular weight fragment was obtained following incubation in urea and 40 000 and 30 000 molecular weight fragments were obtained following heating. By electron microscopy, HA0 rosettes were indistinguishable at neutral pH and pH 5.0, but both heated rosettes and rosettes incubated in Urea formed ball-like structures of similar size from which spike-like structures projected, more distinctly from the urea balls.

Conclusions

(1) Cleaved HA is activated for fusion by incubation at low pH or by heating or incubation in urea at temperatures and concentrations that vary with the strain of virus. The changes in structure that accompany activation are more similar following incubation at low pH or in urea than those observed following heating. These observations agree with previous comparisons of changes in HA structure following heating and low pH incubation (Ruigrok et al 1986, 1988). They do not imply that fusion at neutral pH occurs by a different mechanism, as inferred by Kim and colleagues in their more recent publication (Carr et al 1997).

(2) Uncleaved HA0 is not activated for fusion by incubation at low pH or by heating or by incubation in urea. No changes in HA0 structure are detected at low pH. On the other hand, incubation in urea or heating results in conformational changes that show some similarities, in terms of location, to those displayed by cleaved HA in urea or on heating. However, the temperature and urea

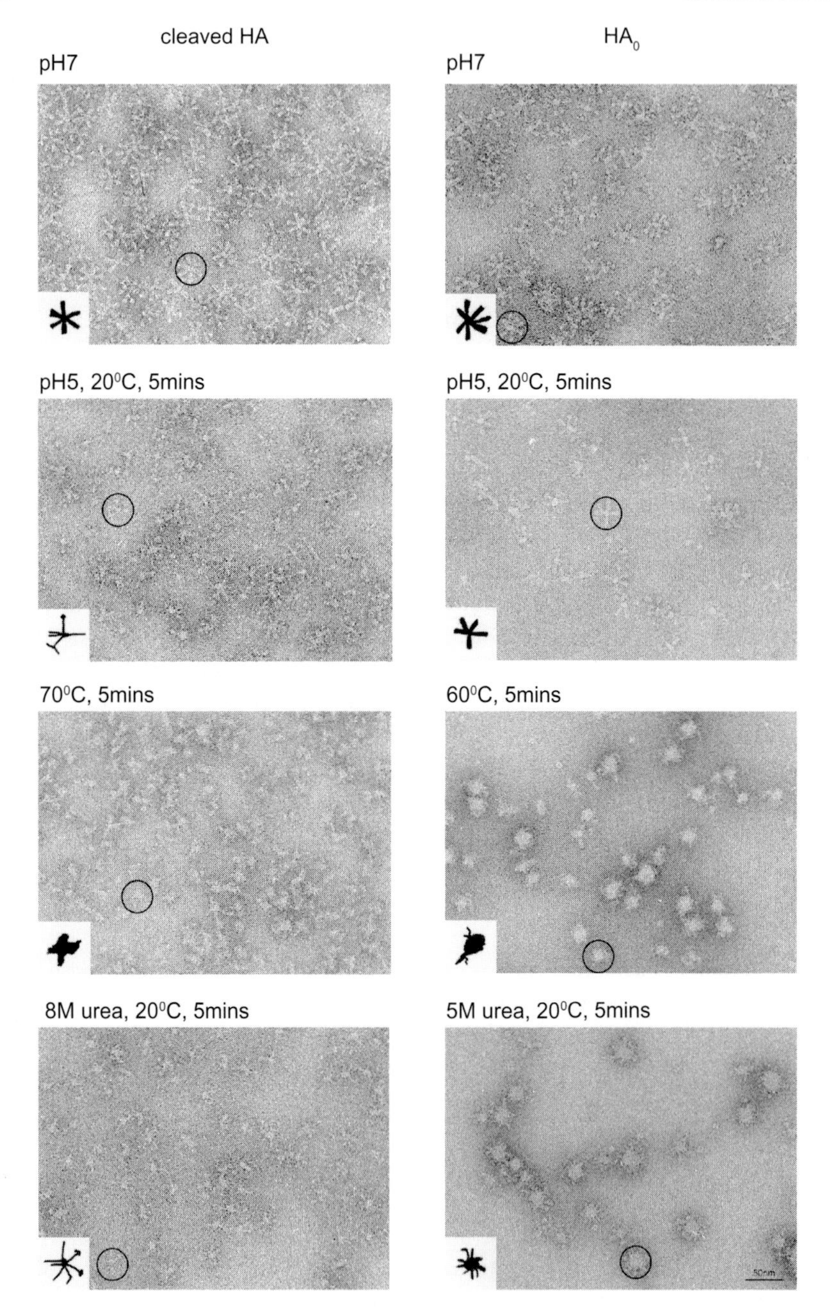

FIG. 3. Electron micrographs of HA0 and HA rosettes incubated at fusion pH, in urea, or heated. The treatment conditions are indicated on each picture together with a diagram of the circled images. Samples were stained in 1% sodium silicotungstate and examined using minimum dose conditions at 100 kV in a Jeol 1200Ex microscope.

TABLE 3 Antigenicity of HA and HA_o under conditions that cause conformational changes

Protein	*Antibody*	*pH*	*Heat*	*Urea*	*No treatment*
HA	HA_1 144	+ve	+ve	+ve	+ve
	HA_2 107	+ve	+ve	+ve	−ve
HA_o	HA_1 144	+ve	+ve	+ve	+ve
	HA_2 107	−ve	+ve	+ve	−ve

The pH of conformational change was measured by a trypsin proteolysis assay (Skehel et al 1982). Temperature and urea concentration estimates were as in Table 1.

concentrations at which these changes in HA0 structure occur are lower than those at which cleaved HA changes conformation. As a consequence, uncleaved HA0 is less stable than cleaved HA suggesting that on cleavage, re-folding of the newly generated 'fusion peptide' N-terminus of HA2 into a negatively charged cavity adjacent to the cleavage site (Chen et al 1998), leads to an increase in the stability of cleaved HA.

(3) The covalent bond that links HA1 to HA2 allows conformational changes in HA0, heated or incubated in urea, which resemble those in cleaved HA when it is similarly treated. However, unlike cleaved HA, the structure of uncleaved HA0 does not change on incubation at low pH. It appears that the response of cleaved HA to low pH requires that the fusion peptide is buried in a cavity of ionizable residues and that as these residues become charged at low pH this location becomes unstable and the 'fusion peptide' is extruded. It seems likely from these comparisons of HA and HA0 that the fusion pH trigger involves these residues exclusively.

References

Bizebard T, Gigant B, Rigolet P et al 1995 Structure of influenza virus haemagglutinin complexed with a neutralizing antibody. Nature 376:92–94

Bullough PA, Hughson FM, Skehel JJ, Wiley DC 1994 Structure of influenza haemagglutinin at the pH of membrane fusion. Nature 371:37–43

Carr CM, Chaudhry C, Kim PS 1997 Influenza hemagglutinin is spring-loaded by a metastable native conformation. Proc Natl Acad Sci USA 94:14306–14313

Chen J, Lee KH, Steinhauer DA, Stevens DJ, Skehel JJ, Wiley DC 1998 Structure of the hemagglutinin precursor cleavage site, a determinant of influenza pathogenicity and the origin of the labile conformation. Cell 95:409–417

Daniels RS, Douglas AR, Skehel JJ, Wiley DC 1983 Analyses of the antigenicity of influenza haemagglutinin at the pH optimum for virus-mediated membrane fusion. J Gen Virol 64:1657–1662

Ruigrok RW, Martin SR, Wharton SA, Skehel JJ, Bayley PM, Wiley DC 1986 Conformational changes in the hemagglutinin of influenza virus which accompany heat-induced fusion of virus with liposomes. Virology 155:484–497

Ruigrok RW, Aitken A, Calder LJ et al 1988 Studies on the structure of the influenza virus haemagglutinin at the pH of membrane fusion. J Gen Virol 69:2785–2795
Skehel JJ, Wiley DC 2000 Receptor binding and membrane fusion in virus entry: the influenza hemagglutinin. Annu Rev Biochem 69:531–569
Skehel JJ, Bayley PM, Brown EB et al 1982 Changes in the conformation of influenza virus hemagglutinin at the pH optimum of virus-mediated membrane fusion. Proc Natl Acad Sci USA 79:968–972
Wharton SA, Skehel JJ, Wiley DC 1986 Studies of influenza haemagglutinin-mediated membrane fusion. Virology 149:27–35
Wilson IA, Skehel JJ, Wiley DC 1981 Structure of the haemagglutinin membrane glycoprotein of influenza virus at 3 Å resolution. Nature 289:366–373

DISCUSSION

Webster: If you look at the high path virus of H5 and H7, how do they alter their structure, or are they both similar?

Skehel: They are in different clades, but I don't think that matters in the sense that H1 and H2 of humans are in the same clade and different from H3. The thing they both have is a different cleavage site. H5 and H7 are classified on the basis that the pathogenic ones have this basic sequence at the site of cleavage. We have the structure of H3 as an uncleaved haemagluttinin (HA), with a single arginine. The site of cleavage is a prominent loop on the surface of the HA. This loop can be covered up by introduction of a glycosylation site just above it, which presumably stops the enzyme getting in. But it can be made bigger by the polybasic cleavage site which the pathogenic H5s and H7s acquire by passage in domesticated birds. There is also work showing that chunks of nucleoprotein can be inserted at the cleavage site (Orlich et al 1994). This goes along with the idea that if this loop is enlarged, it is more easily cleaved. Easier to cleave means more infectious. I think that the majority of human influenza viruses are not that infectious, because it is an adventitious event. There is a protease being secreted at the time in the lung. The pathogenic viruses, therefore produce more infectious virus, which correlates with pathogenicity.

Webster: What about 1918?

Skehel: We haven't done uncleaved 1918. Wilson and colleagues did the structure of the uncleaved HA of 1918 out of insect cells. They don't get the same loop structure as we did for the H3 HA precursor.

Peiris: Following on from the cleavage site issue, why don't we have multibasic cleavage sites and high path viruses emerging in influenza subtypes other than H5 and H7?

Skehel: Presumably, because a single arginine is enough to let a virus get by. Once H5 and H7 go from wild birds to domesticated birds there is slippage of the polymerase and introduction of these polybasic sequences, simply because the single arginine virus isn't growing well enough. It is a mutation to allow growth. Unfortunately, it allows pathogenicity as well. It is interesting that viruses such as RSV, HIV and some bird flus have a polybasic cleavage site. The advantage is that it is cleaved intracellularly by a cellular enzyme that is present ubiquitously.

Lai: Does the cleavage affect the antigenicity of the protein?

Skehel: Not at all: it affects the position of just 19 residues of 550, from residues 323–328 of HA1 and then residues 1–12 of HA2. But if you don't get cleavage you won't have much virus, because cleavage is needed to get to the next cycle of virus replication.

Lai: After cleavage, does fusion of the viral envelope with the cellular membrane require energy?

Skehel: Yes, but where does the energy come from? There are ideas that it might come from the refolding process, which is irreversible. People have the idea that the fusion peptide shoots into the cellular membrane and somehow or other because of the refolding the two membranes are brought together, and this might require energy. For the mechanism, the idea is that the two outer leaflets fuse, and in the process a stalk-like structure is formed. Because the centre of the stalk is hydrophobic, the two inner leaflets collapse.

Lai: Would you predict that it doesn't require energy?

Skehel: There is no obvious linkage to a standard energy-producing system.

Webster: Is inhibition of a fusion peptide a good target? It was suggested by Purnell Choppin years ago.

Skehel: There are compounds that block the conformational change. We have tried to soak these compounds into haemagluttinin to get a structure, without success. It is known from cross-linking studies and selecting mutations that these molecules bind in the region of the fusion peptide. There is one molecule which by cross-linking shows that the link is close to the fusion peptide itself. There are others that cross-link close to residue 106, and then there are experiments selecting mutants, which were at 108 in the central helix. Yes, there have been attempts to block the conformational change, and they have been successful *in vitro* (Cianci et al 1999). There have also been attempts to block cleavage, and they have been successful too. In Germany, Klenk and his colleagues have used ketones (Garten et al 1989), and in Japan a natural inhibitor has been described by Kido's group (Beppu et al 1997), some sort of detergent-extracted material to block proteolysis. This also blocks replication.

Kawaoka: You are saying that cleaved HA is more stable than uncleaved HA. This means that in nature HA with a single basic residue at the cleavage site is more stable. Is that correct?

Skehel: No. There is essentially a linear relationship between pH of cleavage and temperature of cleavage. The lower the pH of cleavage, the more stable it is and the higher temperature needed to melt it. If you are asking about the stability of the HA overall, mutations can make the HA remarkably unstable. Certainly, they can make it sufficiently unstable that the uncleaved HA would be sensitive to around 40 °C. There may be some HA mutants formed which don't make it, because the mutation has rendered them too sensitive to heat. You would imagine that there would be a number of selective pressures. If you have a precursor that is too unstable, this would be a disadvantage. To get a combination of mutations that make it sensitive to temperature would be a selective pressure. You need a precursor that has sufficient stability. But then the next stage is you cleave, you bury the fusion peptide and as a consequence stability increases. You then need something that is sufficiently unstable to be triggerable by low pH. These are obvious selection pressures, but they are different: one is a selection pressure on the stability of the precursor, while the other is a pressure on the response at low pH for fusion activation.

Lai: You mentioned that the polymerase undergoes some sort of slippage, which results in low pathogenecity. Are there certain factors that result in this increased slippage?

Skehel: I think it's a feature of the sequence of the genome of the HA segment at the site of the cleavage loop: a run of adenines. This was shown by Mike Perdue and his colleagues (Garcia et al 1996).

Webster: How easy is to convert an α2,3 to an α2,6 binder?

Skehel: It seems clear that both the H2 and H3 did it by two mutations, in 226 and 228. I don't know how readily this happens. Unfortunately, we don't have them in H5 naturally yet. This doesn't mean to say that the H5 will do it the same way. H1 didn't do it the same way: it didn't have to change at all. The H1s that are circulating now still have the two avian residues in those positions. Of all the HAs we have looked at, H1 has the structurally most adaptable receptor binding site. As a result, we think that the glutamine at 226 is able to fall into the site, move by about an angstrom, and in that position it is not interacting with the glycosidic oxygen or the oxygens of the galactose at position 2. As a consequence, the 2,3-linked receptor can be bound and the 2,6-linked receptor can be bound; if it binds the 2,6-linked receptor the glutamine at 226 drops out of the way; if it binds the 2,3-linked receptor glutamine 226 is higher in the site by about 1 angstrom. This is what we see in the receptor analogue complexes.

Smith: I want to return to your reference to Underwood's comment about antigenic changes in sites A and B having an effect on receptor binding. How would you suggest exploring this? There is a balance between antigenic change in one advantageous mutation and the cost to the virus. This is an interesting balance that might inform cluster transitions.

Skehel: Antigenic changes are one possibility. Influenza B, for example, has an extra glycosylation site around 150. Some of the changes in the receptor binding site in natural isolates are changes of what we would normally consider to be conserved residues of the site. The possibility is that you may change some aspect of receptor binding, such as interactions with the sugars distant from the sialic acid, by changing antigenically. But you might also change some of the residues involved in receptor binding if the receptor binding site itself is put under some pressure, like at the end of a pandemic. There have been changes in things like leucine at 226 to valine. There are some changes which would make sense of being directly influencing receptor binding. As far as I know, no decent receptor binding studies have been done on these viruses.

Smith: What do you think of the glycan array studies, where the different viruses show different binding patterns?

Skehel: I don't think they have made much extra sense for flu yet. They have shown that it binds sialic acid with linkage specificity They have one result showing that binding is particularly good to a bivalent ligand, which doesn't make a lot of biological sense yet. Who knows?

Smith: Do you feel that the substitutions in sites A and B that are a little bit further away from the binding site are mostly free substitutions in terms of their effect on receptor binding, and that they would only change antigenicity?

Skehel: I think they would mainly change antigenicity. Most of them are pointing out into space as far as the HA is concerned. It would be nice to know if there is a preferred position for binding neutralizing antibodies. *In vitro*, the preferred position is right on the tip where there is no restriction. Yet when you look at HA sequences, you find there are changes all over the membrane distal region. You have to imagine that if antibodies contact the lower sites less efficiently these might be less relevant for vaccination, even though they may have just the same relevance for variation.

Holmes: On the adamantine resistance story, one of the mutations that does change, co-inciding with the M2 change, is at 225. This is the diagnostic change in HA1.

Skehel: 225 is right next to 226, obviously! It is more accessible to antibodies than 226. In H1 it participates directly in receptor binding. It is an aspartate that makes hydrogen bonds with Gal-2.

Holmes: The antigenic maps didn't show anything dramatic occurring with these viruses, but there is the change at 225.

Skehel: There are funny things about the locations of mutations that influence things such as pH sensitivity. There are some mutations in that region, of which 218 is a notable example, which can influence pH. If you add a lot of amantadine, you can select for those pH mutants.

Holmes: It is the only variation you see at that site. It suggests to me that it is selected for something. I have a question stemming from an earlier discussion. What mutations are required for an avian virus to spread efficiently in a human population? Do we know the steps?

Skehel: Receptor binding is only part of it. Looking at the H5 site in comparison with H1, you would say that 226 and 228 changes would give the same thing. But we know that there are other mutations around there that seem to be influencing receptor binding specificity *in vitro.*

Holmes: Would this adaptation to human only occur with a reassortment with an existing human H3N2 virus?

Skehel: No. Are you thinking that you can get a virus that has mixed HAs and not reassorted ones?

Holmes: I am more interested in the other segments involved: what other segment mutations are required to do this?

Kawaoka: We know that PB2 is one of them. There may well be others that we don't yet know of.

Holmes: So it is not an insignificant number of changes.

Osterhaus: The point is that once you go over the threshold, you will gradually accumulate it anyway. The HA might be an important factor there.

Holmes: What is the threshold set at, and what tips it?

Skehel: The other issue is whether the human-specific mutations influence polymerase function in the avians. They probably don't since the change in PB2 seems to have cropped up already in avian H5s.

Osterhaus: This is the strange thing. We see this perpetuating in wild birds.

Webster: The pressure of mutations in PB2 in wild birds previously associated with pathogenicity of H5N1 in humans is worrying. Have the wild birds picked up a virus that has been passaged in a mammal?.

Kawaoka: If this is the case, why don't we have other avian viruses with the same change in PB2? I think what has happened is that once the mutation at position 627 is introduced, there must be other mutations in that PB2 that stabilize the 627 mutation

Holmes: Can you map those?

Kawaoka: It is not that straightforward.

Osterhaus: So the compensatory mutations have not been identified.

Webster: Where do these viruses pick up the mammalian trait?

Holmes: It could be happening all the time. It could be that the way evolution works is that the birds are picking up mammalian mutations, and it is surviving because of compensatory changes. The reason why we see it in H5N1 is that for the first time you have sampled thousands of isolates.

Kawaoka: That's not true. If we look at the entire avian virus sequences, only a few avian viruses carry a residue other than glutamic acid. There is a strong selective pressure at that position.

Osterhaus: It could be because the virus has been flipping back and forth between mammalian and avian species so often. It might be that the virus is omnipresent and so has ample opportunity to go back and forth all the time.

Skehel: The business of giving a vaccine before something happens is interesting. There are people who express concerns that by priming in this way, we might influence what response is made when you actually get the infection.

Osterhaus: It is not impossible, but it doesn't look very likely to me. The most practical solution, if we are talking about inactivated vaccine, is that because there are promising adjuvants, we should be stockpiling the adjuvant, providing this gives broad, interspecific cross protection. Rather than stockpiling vaccine which would have a limited life, you could keep a rolling stock of adjuvant and adjust the antigen from time to time.

Skehel: How many companies have such adjuvants?

Osterhaus: There is at least one company that can do it now, and most likely two or three others. The crucial thing is the adjuvant, not the antigen.

Wood: The problem is that to make your antigen will take 3 or 4 months.

Osterhaus: If you just go for prepandemic stockpiling of the vaccine, if you don't have the proper adjuvant this is dangerous. If governments would make a contract with a company to provide antigen within three months, if you have a good adjuvant stockpiled this would be the strategy to go for.

Webster: You have to do both. There is some evidence from the 35 humans primed some years ago in the USA, that if you prime to one of the H5s and come back later you have a distinct advantage in responding to the variants.

Osterhaus: The USA has stockpiled antigen. It is useless now, unless this is tried with some of the adjuvants that are around which result in broad protection.

Webster: We need to do some research on priming.

Wood: We don't know the dose.

Osterhaus: We need to do a dose escalation study in a small number of people.

Wood: We might not need an adjuvant.

Osterhaus: I disagree.

Peiris: To follow your argument about stockpiling adjuvant, do we know that the same GSK adjuvant would do well with H9 haemagluttinin, for example?

Osterhaus: We don't know whether it would protect outside the subtype. The experiments need to be done. If you stockpiled the antigen as you have done in the USA, then you'd better do the experiments with and without a good adjuvant to see if you can prime.

References

Beppu Y, Imamura Y, Tashiro M, Towatari T, Ariga H, Kido H 1997 Human mucus protease inhibitor in airway fluids is a potential defensive compound against infection with influenza A and Sendai viruses. J Biochem (Tokyo) 121:309–311

Cianci C, Yu KL, Dischino DD et al 1999 pH-dependent changes in photoaffinity labelling patterns of the H1 influenza virus hemagglutinin by using an inhibitor of viral fusion. J Virol 3:1785–1794

Garcia M, Crawford JM, Latimer JW, Rivera-Cruz E, Perdue M 1996 Heterogeneity in the hemagglutinin gene and emergence of the highly pathogenic phenotype among recent H5N2 avian influenza viruses. J Gen Virol 7:1493–1504

Garten W, Stieneke A, Shaw E, Wikstrom P, Klenk HD 1989 Inhibition of proteolytic activation of influenza virus hemagglutinin by specific peptidyl chloroalkyl ketones. Virology 172:25–31

Orlich M, Gottwald H, Rott R 1994 Non-homologous recombination between the haemagglutinin gene and the nucleoprotein gene of an influenza virus. Virology 204:462–465

Singapore SARS experience and preparation for future outbreak

Yee Sin Leo

Communicable Disease Centre, Tan Tock Seng Hospital, Moulmein Road, Singapore 308433

Abstract. The SARS outbreak began in Singapore on the 1st March 2003 when the first case was admitted to Tan Tock Seng Hospital. It ended when the last case of the outbreak was isolated on 11th May 2003. A total of 206 probable SARS cases were diagnosed based on World Health Organization (WHO)'s case surveillance definition. A further 32 cases were later added to the total number of 328 cases after virological confirmation amongst the suspected cases. Public health control measures applied along three fronts: prevention and control within the healthcare setting, the community and the borders. The discussions here mainly focus on health care setting, as close to 75% of infected individuals acquired the infection at healthcare institutions. Close to 41% were health-care workers. It is widely accepted that several key factors significantly impact the success in prevention and control measures in a health care setting. These include implementation of a triage system at the health care entry points, swift response and strict adherence of personal protection equipment, hospital temperature surveillance system among health care workers and the designation of a SARS hospital. On the other hand, there were several key factors challenging prevention and control measures. These include the non-specific clinical features in the early disease stage, the less well-understood transmission pattern surrounding super-spreading events, current complex health care delivery leading to frequent patient movement across disciplines and institutions. To adequately prepare for the re-emergence of SARS, collaborative efforts are much needed to enhance early detection, prevention and control preparedness.

2008 Novel and re-emerging respiratory viral diseases. Wiley, Chichester (Novartis Foundation Symposium 290) p 69–78

One of the things that we, the people on the ground, are often asked by policy-makers about the SARS epidemic, is why did we in Singapore publish so little about their SARS experience. If I key in 'SARS Singapore' into PubMed, I have a total of 107 publications. But if I look at the distribution of these papers, we see there are very few on laboratory science. Hopefully, this meeting will bring some of the Singapore researchers together. Many of these papers from Singapore dealt with clinical manifestations of SARS and epidemiological work. Three-quarters of these papers were published in overseas journals.

Internationally, there were around 8000 probable SARS cases, using the WHO case surveillance definitions, which led to 774 deaths. The Singapore story began with two pairs of female companions who travelled to Hong Kong to shop. They booked into the ninth floor of a hotel in Hong Kong, sharing the floor with a professor from Guandong in China. Three of the four became symptomatic, and of these, one happened to be a 'super-spreader' and was admitted to Tan Tock Seng Hospital, where I work. This started the entire SARS epidemic in Singapore.

Our experience has essentially been quite short and sharp. We had a bimodal distribution of cases. The first part was predominantly contributed by Tan Tock Seng hospital, with subsequent spread to the other general hospital, which led to the second part. We have an outlier which is the last case of SARS, and up to this day we still don't know how this patient got SARS.

40% of SARS cases in Singapore were healthcare workers. We had predominantly intrahospital transmissions, with just 17% of transmissions occurring in households. Six cases (2.5%) were undefined, where we couldn't identify the source.

The overall response was that we brought in international help from the WHO, as well as a strong presence in SARS control played by the inter-ministerial committee, where we had seven ministers looking at one disease, which is unprecedented. Within the Ministry of Health (MOH) there was a SARS task force, and three key prevention approaches were implemented. They looked at healthcare institution measures, community measures and international measures. In this paper, I will mainly concentrate on the healthcare institution measures, but I will begin by describing the other measures.

Intensive contract tracing programmes were put in by the MOH, together with expertise from other areas. Because of the spread from patients who had been discharged from hospital, home quarantine measures were introduced. It was only then that we realized the Infectious Disease Act in Singapore has no such legal power, so during the period of this massive response to SARS the act was amended to allow the MOH additional power.

Temperature screening was also implemented, using thermal scanners. In Singapore we have a dedicated ambulance service for SARS cases. At the first point of contact in any healthcare setting, should a clinician suspect SARS, there is a designated number they can call for the special ambulance service with infection control measures in place. We also closed our schools for two weeks, and Hong Kong followed suit.

The first of the healthcare institution measures was the triage system. This was started at the very beginning of the SARS outbreak. In our hospital we have isolation wards. Any of the patients presented to our centre will first go through a questionnaire to assess their symptoms. We use WHO case surveillance definitions, and the symptomatic patients who show fever and respiratory symptoms will be handled in a designated set of consultation rooms, which are isolation facilities. Opposite this, another separate building is used for screening asymptomatic cases. On the basis of our experience with other diseases, when SARS arrived in Singapore we decided to implement a triage system right at the beginning. This concept then transplanted to our parent hospital, Tan Tock Seng, when the Communicable Disease Centre couldn't cope with the patient load. The same principles were used in the Emergency Department (ED): a tent was erected just outside the ED, and this was used for several functions. Our temporary tent was compartmentalized to make consultation rooms, a triage and registration centre, and an X-ray room, with the aim to provide a one-stop service. For those patients who were symptomatic and were required to be separated from the rest, they were handled outside the ED building. The rest of the general cases were handled within the ED building.

The next big challenge came when the cases accumulated. The Communicable Disease Centre (CDC) in Singapore has a different function to the CDCs in other places. We are a patient care facility, not a policy or surveillance service, although we do have a small surveillance unit mainly for HIV. CDC has about 150 beds, and Tan Tock Seng Hospital has about 1000 beds with 150 admissions each day. The ED attends to 300–400 cases each day. In this hospital, 80% of admissions come from the ED. Tan Tock Seng was the first hospital to experience SARS in Singapore, and when CDC was full the issue was whether or not Tan Tock Seng should be used as a designated SARS hospital, or whether the cases should overflow to another hospital. By closing down Tan Tock Seng and making it a designated SARS hospital, the MOH would have to look for another 1000 beds for general cases. It was a big decision at the time. The first option was taken, and Tan Tock Seng was closed for general cases and made a designated SARS hospital.

In the first two weeks of the SARS epidemic we had two of the three imported cases admitted, and the index case, a 'super-spreader', was admitted on 1st March 2003. Singapore only received the international warning from the WHO on 13th March. This has led to a series of responses, including the opening of the command and control centre, and a whole host of infection control measures were implemented in sequential manner. At the end of March, there were no more intrahospital transmission in Tan Tock Seng Hospital except for a tiny blip. There was a price for us to pay when we decided to use the hospital as a designated SARS care site. We utilized the 'no in but out' concept, only bringing in cases with suspected SARS, but allowing patients we didn't think have SARS to be discharged from the hospital. A 60 year-old man was warded in the same ward with the first-index SARS patients. He then became ill after discharge. At the time, when he required admission, he could no longer access Tan Tock Seng Hospital because it had been closed to general admissions. He was admitted to Singapore General Hospital with atypical manifestations, but he had *Escherichia coli* bacteraemia. He was nursed in two open areas that led to a cluster of transmissions in this hospital, with close to 40 immediate secondary cases.

Another measure we put in place was the staff surveillance. Every healthcare worker in the hospital had their temperature monitored three times a day. Should anyone develop fever for two consecutive readings they had to present to the emergency department for further assessment. Options were that these healthcare workers would be put under home quarantine for continued monitoring, or were isolated for further observation. This proved to be very effective as we picked up about a dozen staff with early symptoms.

Contact tracing teams within the healthcare institutions were hard at work. Because of this effort we were able to trace all sources but six of the SARS patients.

Strict infection control practices were employed. We realized the importance of intrahopsital transmissions, so measures were taken to prevent transmissions from patients–patients, patients–visitors, patients–staff and staff–staff. These included compulsory mask fitting tests where the infection control practitioners would test all healthcare workers to make sure they had a well-fitted N95 mask. All healthcare workers put on PPE before they entered the patient care area, and discarded the PPE except for their mask before they came out to the clean area. In addition to this, in ICU we encouraged the staff to use a PAPR for high-risk procedures. There were no data to clearly guide what the best measures at that point were: we just used what was the best practice. One thing we thought we should definitely address was the re-engineering of the working environment in the hospital. We converted the paying class into an isolation facility, and in each single bed facility we put in an industrial suction fan. This provided a unilateral airflow from a clean air supply blowing from the nurses station, to the patient area, then out to the environment. The suction power was strong enough that the door could not be closed properly. In addition, the engineering team augmented an automation process to ensure a safe pathway where contaminated patient areas did not cross with clean areas.

There were also restrictions on the movement of healthcare workers. None of us were allowed to cross cover other hospitals. There were also restrictions on the movements of visitors, who had to register before they entered clinical areas, and go through a thermal scanner.

Lastly, I want to mention the effort of the MOH in terms of communication, which is one of the key elements of the control and prevention work. We had daily meetings at the MOH. Patients' information was posted on the SARS web, and this allowed different institutions to access information on movement of patients, and whether or not they were in the suspect group, so they could isolate them before any confirmation of the disease state. Moving forwards, I am sure that researchers such as many in this group have a better understanding of the human coronavirus. This information will be key in our prevention efforts, and in our planning. Is there a reservoir for human SARS coronavirus? We don't have a clear understanding of this. The mode of transmission we now know to be mainly contact and droplet transmission. Information on the incubation period and period of transmissibility are available, and we do have a susceptible population.

The Singapore data show that most cases didn't generate any secondary cases. But some individuals were 'super-spreaders', and resulted in as many as 40 secondary direct transmissions. There were five super-spreading events. Out of these five, there were three atypical cases. One, which I mentioned earlier, had very few respiratory symptoms but significant gastro-intestinal symptoms. His chest X-ray changes were unremarkable so that no one suspected SARS. We had another super-spreading event involving an elderly lady with multiple co-morbidites. She had multiple risk factors for fever and pneumonia; SARS was not suspected and that gave rise to a super-spreading event in the coronary care unit. The third was the brother of the super-spreader mentioned who spread the disease in his workplace as well as causing a small cluster in the National University Hospital.

We still don't understand why these super-spreading individuals had such a high ability to transmit SARS. The only two factors in common in the Singapore cohort were (1) delay to isolation of five days or more and (2) admissions to a non-isolation ward with shared common facilities. Beyond that we can't find any good indicators to differentiate at onset super-spreaders from non-spreaders. If we look at all the cases with secondary transmissions, and look at days to isolation after onset, it tends to be 5 days or beyond. This corresponds nicely with Malik Peiris' data on secretions of the virus in the respiratory tract. This tends to be more significant in the later part of the disease. However, there are some individuals who are in non-isolation facilities for long periods, yet they fail to spread the disease.

We looked at SARS attack rate in healthcare facilities. We had two wards in Tan Tock Seng Hospital that experienced SARS transmissions before we were informed of the atypical pneumonia outbreak, and we had 56% attack rate. This is extremely high. In the non-isolation facilities, half of the healthcare workers in these locations were infected. Most presented with pneumonia, and we also picked up a number of healthcare workers who were asymptomatic. These data are different from many of the other publications, where asymptomatic SARS is extremely rare.

To try to understand transmissions, we did a case control study looking at healthcare workers with exposure to SARS patients within the same unit. We examined the personal protection equipment they used. As part of the infection control, from the beginning we used N95 masks. Those using masks were significantly better protected, which shows that droplet transmission is involved. Those who practised good hand washing also had a significantly lower attack rate. Other studies from Hong Kong have shown that even surgical masks confer some protection, but N95 masks are better.

Some of the problems remaining with SARS include our poor understanding of super-spreading events, the non-specific nature of the early clinical symptoms, and our inability to make an early diagnosis. For this reason, to control SARS we still need a broad-based surveillance, together with broad-based isolation and quarantine techniques. Looking forward, in terms of prevention and control there are still many measures that we need to put in place. Is there a specific treatment for SARS at this point? We did use ribovarin for some cases in Singapore. But there is effectively no clinical trial that has fully addressed the best treatment approach. Would animal data be useful? Could α interferons be used in clinical cases on the basis of results in animal models? Possibly. We need to consider this. I don't know how soon we could have a vaccine for SARS if it were to reappear. Finally, communication is certainly one of the keys in our combat against SARS.

DISCUSSION

Webster: If you had to do it over again, what are the key issues that we can learn from you for combating future emerging infections?

Leo: I thought I would restrict this to SARS, which is spread by droplet and direct transmission. It is very different from the response we would make if we were confronted by influenza or other pathogens with a different mode of transmission. There are still many things that are not understood about SARS. Because of this knowledge gap, I am an advocate for a broad-based response knowing that there are 'superspreaders' that spread more than others. There is no clear indication to differentiate a 'superspreader' event from individuals who don't spread. It would appear practical to rely on day 1 or 2 of a febrile episode as an indicator to determine symptoms for isolation and quarantine. Beyond that, the research community has to work on early diagnosis as well as development of vaccines in anticipation of a big community outbreak. Using laboratory markers would also be important for clinicians to work together towards some kind of clinical protocol that we could use.

Webster: Did you say that regular surgical masks are effective?

Leo: That was the Hong Kong study, but it's not our experience. When we implemented infection control we used N95 straight away. The Hong Kong clinicians reported their experience is that the surgical masks gave significant protection.

Tambayh: They gave the same protection as N95 masks. Both were better than paper masks (Seto et al 2003).

Osterhaus: Is it the masks or the psychology behind them?

Webster: What measures have you put in place in Singapore as a result of SARS? Do you have a huge stash of N95s and PAPRs (powered air purifying respirators)?

Leo: The ministry has put a lot of effort into stockpiling masks and personal protective equipment, as well as PAPRs for staff protection. Most wards would now have at least one or two PAPRs. The stockpile is centralized at the ministry.

Ling: We also have a stockpile of lab kits for SARS diagnosis.

Su: The key knowledge that came from Taiwan is that if there is no fever, there is no transmission. This helped us to control SARS as we could easily screen patients.

Webster: Wouldn't that depend on the status of the patient? When there is no fever, what stage of infection is it?

Su: Fever is seen after 2–3 days. This kind of knowledge is very important for public health control.

Peiris: You talked about the decision to close down Tan Tock Seng Hospital for general admissions, and you had to find beds elsewhere. What happened to overall hospital admissions during that time? The demand probably came down, didn't it?

Leo: It depends on the individual hospital setting. For Tan Tock Seng Hospital, admissions mainly came from emergency departments, so we were functioning very much like an acute hospital. There are certain hospitals where if the elective load is 50% of the hospital admissions, then their elective admissions can be cut back. The rest of the hospitals would then have to take on the acute hospital load from Tan Tock Seng Hospital, and this is what happened during SARS. If we look at the national level we saw a trend during the SARS period that there was a fall in use of all the health facilities. Possibly psychology also plays a part, as well: the hospital is now viewed as a risk area, and people don't want to enter it unless it is absolutely necessary. It took us three months to see a rebound.

Ling: One other thing we should note is that we broke our clinicians into two teams, and one stood down while the other worked. They were afraid that if a lot of doctors got sick we would have no services left.

Leo: This wasn't the case in Tan Tock Seng Hospital. All health workers carried on working and we used staff temperature monitoring as indicators for whether or not the staff were infected. Transmission in Tan Tock Seng Hospital ceased after staff used PPG in all patient care areas. We did experience an unusual peak at the end, where we believe because of smouldering transmissions in one particular ward, there was a demented elderly patient who had multiple episodes of other medical illness and there was no way we could differentiate clearly whether there was any episodes of SARS because of these ongoing medical conditions. In this ward we experienced transmission of three more health care workers and thought that it was because of this disturbed patient, who grabbed the facemasks from healthcare workers, together with extensive physical contact during the course of nursing.

Hoffmann: I have a question with regard to transmission and the super spreaders. Is the sequence of viruses isolated from super spreaders known?

Ling: The sequences are the same. There was no real difference between the super spreader and non-super spreader virus sequences.

Hoffmann: So it is the host then?

Ling: We don't know.

Peiris: There was nothing special about the virus that we can observe. But it may not be just the host or the virus. It is very likely the whole context. If we look at super spreading events outside the hospital, for example, that at Hotel M and at Amoy Gardens in Hong Kong, in each case there was a unique constellation of environmental factors at play. It is the overall situation taken together that has a major contribution to the super spreading.

Lai: One issue is whether asymptomatic cases can spread the disease or not. You had a few cases of asymptomatic SARS: was there any evidence that these patients spread the disease?

Leo: For the asymptomatic patients we went back and asked for the history. There was no spread as far as we could tell. I don't know whether or not we can safely conclude that without fever there is no transmission. We do have some atypical cases. Some have temperature below 38.5 for one or two days, and subsequently serology proved to have seroconversion.

Ling: I think I remember seeing temperatures of 37.

Osterhaus: With the current knowledge of epidemiology, knowing that a lot of work is ongoing on vaccine and antiviral development, would it be sufficient to put a protocol together to implement conservative without stockpiling any vaccines or medication whatsoever. This is what we are doing: is it sufficient, or do we need to continue developing a vaccine and monoclonal antibodies? Would normal containing measures be sufficient for SARS? We are facing this situation. The reservoir is predominantly in some laboratories, but there might be an issue of reintroduction.

Peiris: If a virus similar to the SARS coronavirus that was present in 2003 re-emerges, I think we can control it using public health measures. It would be great to have antivirals and vaccines, but it is probably not essential. However, if the virus that re-emerges has changed in its transmission dynamics, with transmission earlier in the course of the disease, then all bets are off.

Lai: Why did SARS disappear? Can we credit this entirely to the success of public health measures?

Osterhaus: Don't forget the diagnostics. Having diagnostic methods helped a lot.

Lai: Retrospectively, can we say that the success in control of SARS was due to public health measures?

Webster: Yes, and the infrastructure that permitted rapid identification.

Osterhaus: The identification of patients who might be shedding is crucial. Putting people in isolation or quarantine is essential. It would not have been possible to do this if it wasn't possible to discriminate between SARS and other acute respiratory virus infections.

Anderson: It is clear that control measures were a key part of the control of SARS. Other factors such as seasonality and risk of transmission may have contributed. Identifying cases, identifying contacts and managing contacts are important. Some of this is quarantine, but it doesn't necessarily involve quarantine. Identification of contacts is one of the things that we learned was very important. For example, a hospitalised patient whose exposure was missed, was not monitored for any SARS-like illness, and SARS illness was missed because of an atypical presentation, infected a second cohort of patients in a hospital in Canada. In retrospect, in a hospital with nosocomial transmission, you should use a very liberal definition of exposure to decrease the risk of missing exposures. Some investigators from Canada suggested that they observed an increase in morbidity and mortality with other illnesses because they didn't have access to clinical to care because of SARS-related restrictions in health care facilities. Is there any suggestion that in Singapore there was more morbidity and mortality from non-SARS illness because of decreased access to healthcare facilities?

Lai: In Taiwan there was a reduction in hospital admissions, but no increase in overall mortality rates during the SARS period. The issue I am interested in is whether public health measures are sufficient to control a future outbreak of SARS.

Anderson: As Malik Peiris said, if it is similar to the SARS that caused the outbreak in 2003, then I think the answer is yes. Very few of us would have predicted that control of a respiratory-spread virus would be achieved with classic public health control measures, but it was. Good case identification, contact tracing and management controlled the spread. In contrast, missed cases were often associated with further transmission. Characteristics of SARS that helped control spread include very few asymptomatic or atypical illnesses, and little transmission early in the illness before the patients became sufficiently ill to seek medical attention or be hospitalized. Many features of transmission are similar to those of smallpox.

Ling: Did you get a lot of coughing and sneezing in your patients? We didn't see that much. We would see this with flu, so we were lucky with SARS because it wasn't quite as transmissible as flu.

Osterhaus: Is there coughing and sneezing with H5N1?

Peiris: There is coughing but not much sneezing in human H5N1 disease. The upper respiratory component was less prominent in both SARS and H5N1 disease.

Leo: It is not just an effect of the absolute numbers we are looking at with SARS; it is also the impact of disease on healthcare workers. There was a period when ICUs were completely full because of SARS. Many of the survivors survived through one or two months of intensive care. This puts a lot of stress on the infrastructure. I think we should continue to try to understand the pathogens more and to look into various treatment modalities. Monoclonal antibodies could be one route.

Peiris: I was not saying that we shouldn't look for treatments or vaccines. The question that was raised was whether public health measures could control a future outbreak, and the answer was yes. It doesn't mean we should stop doing research on SARS.

Osterhaus: We don't know what we are going to be dealing with in the future. Even if it is a SARS-like coronavirus, it might behave differently in terms of transmission.

Holmes: An example is Hendra virus, which in Malaysia didn't get going in humans, but in Bangladesh it has. It is difficult to generalize from what happened a few years ago because the viruses will be evolving in the reservoir.

Osterhaus: Again, it comes down to proper animal surveillance.

Reference

Seto WH, Tsang D, Yung RW et al 2003 Advisors of Expert SARS group of Hospital Authority. Effectiveness of precautions against droplets and contact in prevention of nosocomial transmission of severe acute respiratory syndrome (SARS). Lancet 361:1519–1520

SARS lessons for a young virology laboratory in Singapore

Yee-Joo Tan

Collaborative Antiviral Research Group, Institute of Molecular and Cell Biology, 61 Biopolis Drive, Proteos, Singapore 138673

Abstract. In April 2003, the Collaborative Antiviral Research group started to work on the SARS coronavirus (SARS-CoV). Firstly, we profiled the antibody responses against SARS-CoV proteins during infection. In 74 convalescent sera obtained from confirmed SARS cases in Singapore, we found that all of them contained antibodies against the major structural proteins, nucleocapsid and spike (S), and 70% contained antibodies against one of the accessory proteins termed 3a. The N-terminus of the 3a protein, which is also incorporated into the virion, elicits antibodies that can inhibit the replication of the SARS-CoV in Vero E6, a green African monkey cell-line. A panel of rabbit polyclonal antibodies targeting the S protein was also tested for viral neutralizing activities and it was found that the amino acids 1029–1192 of S contain neutralizing epitopes. Four groups of mouse monoclonal antibodies were also produced using this region of S and they were found to have viral neutralizing activities. By using an *in vitro* cell–cell fusion assay, it was demonstrated that these anti-S antibodies can block membrane fusion, suggesting that the mechanism of inhibition is likely to be through the blocking of viral–cell or cell–cell fusion during the SARS-CoV infection.

2008 Novel and re-emerging respiratory viral diseases. Wiley, Chichester (Novartis Foundation Symposium 290) p 79–88

The Collaborative Antiviral Research (CAVR) group in the Institute of Molecular and Cell Biology (IMCB) was established in 1999 to work on the Hepatitis C virus. Just before the severe acute respiratory syndrome (SARS) outbreak, CAVR had two co-principal investigators, one assistant professor and three post-doctoral fellows. In April 2003, the laboratory was called upon by the acting director of the IMCB (Professor Wanjin Hong), who was also a co-principal investigator of the CAVR at that time, to work on the SARS coronavirus (SARS-CoV). Other members of IMCB were also roped in to help to generate reagents that could be useful for the development of diagnostics assays and anti-viral therapeutics, as well as for basic research. Here, I will present some of the results we have obtained from our research on one of the major SARS-CoV structural proteins, spike (S), and an accessory protein termed 3a.

Profile of antibody responses in SARS patients

In the first project, we expressed various SARS-CoV proteins and used them to detect specific antibodies in convalescent sera obtained from confirmed SARS patients in Singapore. We showed that these sera contained antibodies against the main structural proteins nucleocapsid (N) and spike (S), as well as the 3a protein, which is unique to the SARS coronavirus (Tan et al 2004a). Whereas all the 74 sera tested showed reactivity to N and S, only ~70% of them showed reactivity to 3a. The knowledge gained from this study was used to develop one of the first rapid immunochromatographic assays for detection of SARS infection as well as an ELISA kit (Guan et al 2004). It is interesting that a high percentage of SARS convalescent patients had antibodies against the 3a protein, as the 3a protein is an accessory protein that is not essential for viral replication in cell culture or young mice (Yount et al 2005). Indeed, anti-3a antibodies were found in several cohorts of patients from other countries (Guo et al 2004, Yu et al 2004, Zeng et al 2004) Interestingly, in two separate cohorts of SARS patients, one from Taiwan (Liu et al 2004) and one from Hong Kong (Zhong et al 2005), B cells recognizing the N-terminal region of 3a were isolated from patients.

Characterization of the SARS-CoV accessory protein 3a

3a is a novel coronavirus structural protein (Ito et al 2005, Shen et al 2005), which is also expressed on the cell surface of infected cells (Tan et al 2004b, Ito et al 2005). The topology of 3a on the cell surface was determined experimentally: its first 34 amino acids (aa) (i.e. before the first transmembrane domain), are facing the extracellular matrix and its C-terminus after the third transmembrane domain (i.e. 134–274 aa) is facing the cytoplasm (Fig. 1A; Tan et al 2004b). Accordingly, the N-terminal ectodomain would be expected to protrude out of the virion. We have also generated two rabbit polyclonal antibodies that are targeted to the N- or C-terminus of 3a. As expected, the anti-3a N-terminus antibody could recognize the protein expressed on the cell surface and intracellularly (Fig. 1B), while the anti-3a C-terminus antibody could only recognize the protein expressed intracellularly. Although both antibodies can bind the 3a protein (Fig. 1B, C), only the anti-3a N-terminus antibody can inhibit SARS-CoV propagation in Vero E6 culture (Akerstrom et al 2006).

In another study, it was reported that 48.8% of patients who recovered from SARS had antibodies against the N-terminus of 3a while only 7.4% of the diseased patients had such antibodies (Zhong et al 2006). It was further demonstrated that anti-3a antibodies in the patient serum could bind cells expressing 3a and induce the elimination of these cells in the presence of the human complement system

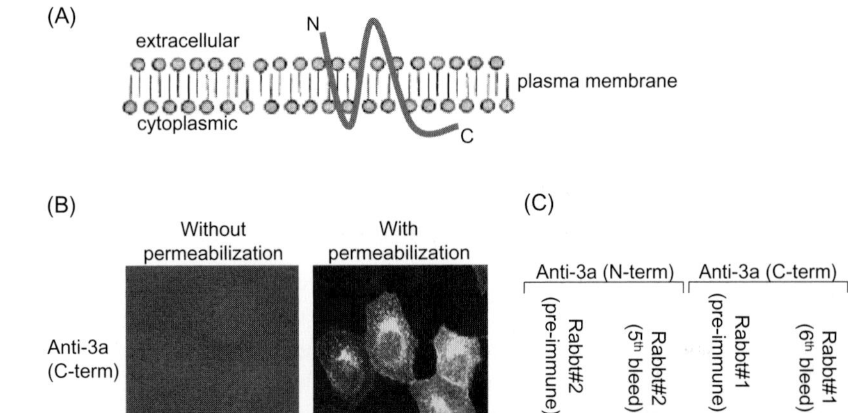

FIG. 1. (A) A schematic diagram showing the topology of the SARS-CoV 3a protein on the plasma membrane. The N-terminus of 3a is protruding into the extracellular matrix while the C-terminus is facing the cytoplasm. The 3a protein spans the plasma membrane three times, i.e. it has three transmembrane domains. (B) Vero E6 cells transiently transfected with a cDNA construct for expressing the SARS-CoV 3a protein were stained with either an anti-3a rabbit polyclonal antibody (upper panel) that recognizes the C-terminus of the protein or an anti-3a rabbit polyclonal antibody (lower panel) that recognizes the N-terminus of the protein. Cells were fixed with 4% paraformaldehyde and used directly (without permeabilization) or permeabilized with Triton X before incubation with antibodies. (C) Vero E6 cells were infected with SARS-CoV at a multiplicity of infection of 1 and then prepared for western blot analysis. Pre-immune serum from the rabbit that was immunized with the 3a N-terminal peptide showed no reactivity (lane 1), while the 5th bleed from the same rabbit after immunization detected the 3a protein in infected cells (lane 2). Similarly, pre-immune serum from the rabbit that was immunized with the 3a C-terminal bacterially expressed protein showed no reactivity (lane 3), while the 6th bleed from the same rabbit after immunization detected the 3a protein in infected cells (lane 4).

(Zhong et al 2006). Taken together, these data suggest that the 3a protein can stimulate protective humoral responses during SARS infection. Even though the 3a protein is not essential for SARS-CoV replication in cell culture and young mice (Yount et al 2005), it is likely that 3a contributes to viral replication or pathogenesis in the natural host(s). Indeed, the deletion of the 3a gene by reverse genetics (Yount et al 2005) or knockdown of 3a expression by RNA interference (Lu et al

2006, Akerstrom et al 2007) causes a decrease in viral replication. It has also been shown that 3a can form an ion channel and may be involved in promoting viral release (Lu et al 2006). In future studies, it will be crucial to determine the exact role of 3a during SARS-CoV infection by using animal models such as aged Balb/c mice, hamsters, ferrets and non-human primates.

Identification of neutralizing epitopes in the SARS-CoV spike (S) protein

Next, we obtained rabbit polyclonal antibodies against five denatured recombinant S protein fragments expressed in *Escherichia coli* and tested them for their abilities to inhibit SARS-CoV propagation in Vero E6 culture (Fig. 2). Our results showed the anti-SΔ10 antibody, which was raised against aa 1029–1192 of S and included the heptad repeat 2 domain (HR2), has neutralizing activities (Keng et al 2005). The interaction between the two heptad repeat domains, HR1 and HR2, in the C-terminal region of S brings the fusion peptide, predicted to be near the N terminus of HR1, into close proximity with the transmembrane domain, and this facilitates the fusion between viral and cellular membranes, allowing the virus to enter the

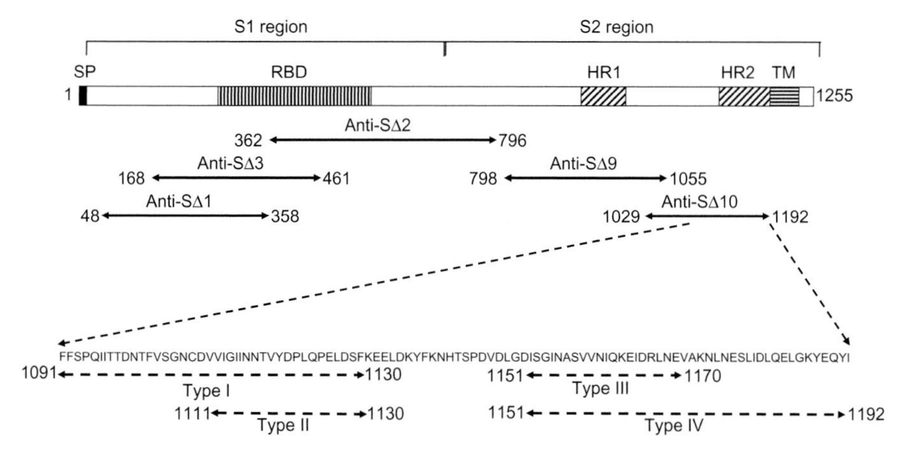

FIG. 2. A schematic diagram showing the different regions of the SARS-CoV spike (S) protein targeted by the rabbit polyclonal and mouse monoclonal antibodies generated in the Collaborative Aantiviral Research group, Institute of Molecular and Cell Biology, Singapore. The regions covered by the solid arrows represent the five bacterially-expressed proteins used to raise the respective polyclonal antibodies (Anti-SΔ1, 2, 3, 9 and 10). The regions covered by the dashed arrows represent the region targeted by the respective mouse monoclonal antibodies (Types I, II, III and IV). Functional motifs within the S protein are shown as boxes: black solid box, signal peptide (SP); box with vertical lines, receptor binding domain (RBD); boxes with slanted lines, heptad repeat domains 1 and 2 (HR1 and HR2); box with horizontal lines, transmembrane domain (TM).

cell. Thus, as would be expected, the anti-SΔ10 antibody could prevent S-mediated membrane fusion in an *in vitro* cell–cell fusion assay (Lip et al 2006). Recently, we have also refined this fusion assay such that secreted alkaline phosphatase (SEAP), which can be quantified accurately, is secreted into the culture supernatant only upon cell–cell fusion (Chou et al 2007). Our results are also consistent with the findings of several other laboratories that peptides from the HR2 region can block SARS-CoV infection (Bosch et al 2004, Ingallinella et al 2004, Yuan et al 2004). As bacterially expressed proteins would be easy and cost-effective to produce on a large scale, the SΔ10 fragment (aa 1029–1192) identified in this study may be an ideal vaccine candidate for SARS-CoV.

In collaboration with the Monoclonal Antibody Unit at IMCB, we also obtained monoclonal antibodies targeting this region of S (Lip et al 2006). This panel of S monoclonal antibodies binds to four distinct regions namely, 1090–1130aa (Type I), 1111–1130aa (Type II), 1151–1170aa (Type III) and 1151–1192aa (Type IV) (Fig. 2). All the monoclonal antibodies have viral neutralizing activities and can block membrane fusion. Interestingly, while the binding domains for the type III and IV antibodies includes the HR2 domain, the binding domains for the type I and II antibodies are found between the HR1 and HR2 motifs and immediately upstream of the HR2 domain. These results suggest that the spacer region between HR1 and HR2 of the SARS-CoV S protein is also important for mediating membrane fusion. Unlike the spike protein of the mouse hepatitis coronavirus, the S protein of SARS-CoV is not cleaved by a protease. However, by sequence comparison, the SARS-CoV S protein may be divided into two regions, S1 (aa 1–690) and S2 (aa 691–1255) (Fig. 2). Numerous laboratories have obtained neutralizing S monoclonal antibodies that bind to the S1 region, which contains the receptor binding domain (see review by Tsunetsugu-Yokota et al 2006). As would be expected, these monoclonal antibodies have high neutralizing titres. However, besides our study, there are only two other reports that described neutralizing antibodies that mapped to the S2 region (Duan et al 2005, Coughlin et al 2007). One of them was shown to bind to aa 1023–1189 and was isolated from B-cells obtained from SARS convalescent patients in China (Duan et al 2005). Recently, the three-dimensional structure of the complex between the receptor binding domain of S and a potent neutralizing antibody (80R) that binds to the S1 region was solved and this reveals the basis of the broad neutralizing ability of the antibody (Hwang et al 2006). In future studies, we would like to perform structural studies on the S2 neutralizing monoclonal antibodies that we have generated, as this may provide important insights into the S-mediated membrane fusion process.

Currently, we are applying the techniques we have learnt in SARS research to the H5N1 influenza A virus. In collaboration with Dr Sunil Lal, India, we have obtained cDNA encoding the viral proteins of a H5N1 isolate that was isolated in

Vietnam in 2004. Our preliminary results show that we have successfully obtained monoclonal antibodies against the HA protein.

References

Akerstrom S, Tan Y-J, Mirazimi A 2006 Amino acids 15 to 28 in the ectodomain of SARS coronavirus 3a protein induces neutralizing antibodies. FEBS Lett 580:3799–3803

Akerstrom S, Mirazimi, A, Tan Y-J 2007 Inhibition of SARS-CoV replication cycle by small interference RNAs silencing specific SARS proteins, 7a/7b, 3a/3b and S. Antiviral Res 73:219–227

Bosch BJ, Martina BEE, Van Der Zee R et al 2004 Severe acute respiratory syndrome coronavirus (SARS-CoV) infection inhibition using spike protein heptad repeat-derived peptides. Proc Natl Acad Sci USA 101:8455–8460

Chou CF, Shen S, Mahadevappa G, Lim SG, Hong W, Tan Y-J 2007 The use of HCV NS3/4A and SEAP to quantitate cell-cell membrane fusion mediated by SARS-CoV S protein and the receptor ACE2. Anal Biochem 366:190–196

Coughlin M, Lou G, Martinez O 2007 Generation and characterization of human monoclonal neutralizing antibodies with distinct binding and sequence features against SARS coronavirus using XenoMouse. Virology 36:93–102

Duan J, Yan X, Guo X et al 2005 A human SARSCoV neutralizing antibody against epitope on S2 protein. Biochem Biophys Res Commun 333:186–193

Guan M, Chen HY, Foo SY, Tan Y-J, Goh PY, Wee SH 2004 Recombinant protein-based enzyme-linked immunosorbent assay and immunochromatographic tests for detection of immunoglobulin G antibodies to severe acute respiratory syndrome (SARS) coronavirus in SARS patients. Clin Diagn Lab Immunol 11:287–291

Guo JP, Petric M, Campbel W, McGeer PL 2004 SARS coronavirus peptides recognized by antibodies in the sera of convalescent cases. Virology 324:251–256

Hwang WC, Lin Y, Santelli E et al 2006 Structural basis of neutralization by a human anti-severe acute respiratory syndrome spike protein antibody, 80R. J Biol Chem 281:34610–34616

Ingallinella P, Bianchi E, Finotto M 2004 Structural characterization of the fusion-active complex of severe acute respiratory syndrome (SARS) coronavirus. Proc Natl Acad Sci USA 101:8709–8714

Ito N, Mossel EC, Narayanan K et al 2005 Severe acute respiratory syndrome coronavirus 3a protein is a viral structural protein. J Virol 79:3182–3186

Keng C-T, Zhang A, Shen S et al 2005 Amino acids 1055 to 1192 in the S2 region of SARS coronavirus S protein induces neutralizing antibodies: implications for the development of vaccine and anti-viral agent. J Virol 79:3289–3296

Lip K-M., Shen S, Yang X et al 2006 Monoclonal antibodies targeting the HR2 domain and the region immediately upstream of the HR2 of the S protein neutralize *in vitro* infection of severe acute respiratory syndrome coronavirus. J Virol 80:941–950

Liu IJ, Hsueh PR, Lin CT et al 2004 Disease-specific B cell epitopes for serum antibodies from patients with severe acute respiratory syndrome (SARS) and serologic detection of SARS antibodies by epitope-based peptide antigens. J Infect Dis 190:797–809

Lu W, Zheng BJ, Xu K et al 2006 Severe acute respiratory syndrome-associated coronavirus 3a protein forms an ion channel and modulates virus release. Proc Natl Acad Sci USA 103:12540–12545

Shen S, Lin PS, Chao YC et al 2005 The severe acute respiratory syndrome coronavirus 3a is a novel structural protein. Biochem Biophys Res Commun 330:286–292

Tan Y-J, Goh P-Y, Fielding BC et al 2004a Profile of antibody responses against SARS-coronavirus recombinant proteins and their potential use as diagnostic markers. Clin Diag Lab Immunol 11:362–371

Tan Y-J, Teng E, Shen S, 2004b A novel SARS coronavirus protein, U274, is transported to the cell surface and undergoes endocytosis. J Virol 78:6723–6734

Tsunetsugu-Yokota Y, Ohnishi K, Takemori T 2006 Severe acute respiratory syndrome (SARS) coronavirus: application of monoclonal antibodies and development of an effective vaccine. Rev Med Virol 16:117–1131

Yount B, Roberts RS, Sims AC et al 2005 Severe acute respiratory syndrome coronavirus group-specific open reading frames encode nonessential functions for replication in cell cultures and mice. J Virol 79:14909–14922

Yu C-J, Chen Y-C, Hsiao C-H et al 2004 Identification of a novel protein 3a from severe acute respiratory syndrome coronavirus FEBS Lett 565:111–116

Yuan K, Yi L, Chen J et al 2004 Suppression of SARS-CoV entry by peptides corresponding to heptad regions on spike glycoprotein. Biochem Biophys Res Commun 319:746–752

Zeng R, Yang RF, Shi MD et al 2004 Characterization of the 3a protein of SARS-associated coronavirus in infected vero E6 cells and SARS patients. J Mol Biol 341:271–279

Zhong X, Yang H, Guo ZF et al 2005 B-cell responses in patients who have recovered from severe acute respiratory syndrome target a dominant site in the S2 domain of the surface spike glycoprotein. J Virol 79:3401–3408

Zhong X, Guo Z, Yang H et al 2006 Amino terminal of the SARS coronavirus protein 3a elicits strong, potentially protective humoral responses in infected patients. J Gen Virol 87:369–373

DISCUSSION

Skehel: Did you isolate any antigenic variants using the anti-S monoclonals? From the point of view of defining the binding site precisely, it would be useful to sequence variants selected by the monoclonal.

Tan: No, we haven't done this. I think the S2 domain is very well conserved.

Lai: It has been done with the murine coronavirus. It is interesting that here the neutralization epitopes are often located in S2. This is apparently also true for SARS virus. There are neutralization epitopes located near the C-termini of S2.

Skehel: The question is, where they are structurally? In HA2 you might imagine that the location of HA2 is not accessible to antibodies, but you don't know in the case of the S protein whether these regions are near the surface or not.

Peiris: Were there attempts to generate escape mutants with the murine coronavirus? Is this possible?

Lai: Yes.

Tan: For our study, we used a denatured form of the S2 fragment. Perhaps this is why the epitopes are exposed. The other study from China (Duan et al 2005) used a B cell library derived from SARS patients, and they found the same epitope.

Skehel: You are going to get antibodies against these peptides, but whether you get neutralizing ones is unclear.

Vasudevan: You showed an assay with cell fusion and syncitium formation. Is syncitium formation in infected cell culture a relevant assay for viral cell membrane fusion?

Skehel: It is not clear what the activator is for fusion activity.

Tan: In this assay it is trypsin.

Vasudevan: In your case, John Skehel, it is pH.

Skehel: Yes, but there are other viruses where the activator is unknown. In RSV it is not clear what the activator is. It is probably cleavage, as it seems to be in this case.

Vasudevan: In dengue there is a dimer to trimer transition of the E protein based on the two structures that have been solved in Steve Harrison's lab. It is hard to imagine this happening on the cell surface of infected vero cells.

Skehel: The activator for dengue is also unknown.

Lai: Heating can also activate the conformational change of the spike protein.

Skehel: Whether this is the physiological activator is unclear.

Peiris: You said that the patients who died didn't have the 3A antibodies, but the patients who survived did. Is this just a time-related finding? Is it a late-developing antibody so the patients who died didn't have time to develop it? Have you looked at the kinetics of the antibody response?

Tan: A group in China have done this (Zhong et al 2006). We don't have the samples in Singapore to do that work.

Lal: This raises one more interesting aspect of 3a. You were able to protect 70% of patients with 3a protein. Then you said that the 3a knockouts were showing a decrease in viral replication. This is an interesting application, because we have recently shown that 3a protein interacts with the 5' UTR of the genome of this virus. It does have some role to play in the initiation of the replication complex that forms.

Kahn: You mentioned that 3a might be an ion channel. Is there any biochemical evidence to support that? You also mentioned that other coronaviruses don't have this function. Have you tried looking at ion channel blockers that could potentially inhibit the function of this protein and decrease its activity?

Tan: The ion channel activity of 3a has been demonstrated biochemically (Lu et al 2006). Whether 3a has a homologue in other coronaviruses is unclear. There was a paper by Peter Rottier's group in the Netherlands (Oostra et al 2006) showing that there are other coronavirus proteins with a similar topology, although the sequence identity is not high.

Skehel: What is the suggested function?

Tan: It is virus release. But we don't completely understand how it works. It is not essential for viral replication in cell culture or young mice (Yount et al 2005).

Lai: Ion channels have not been found in other coronaviruses.

Liu: The E proteins can also form an ion channel, detected using patch clamp. This was suggested to be involved in virus release and also budding. Potentially, the E proteins could form an ion channel on the Golgi membrane.

Kahn: Have you tried any ion channel blockers?

Liu: Yes. Amantadine blocks replication of SARS.

Skehel: Is this using high concentrations? It's important.

Liu: I don't recall.

Skehel: The M2 effect in the Golgi is definitely at low concentrations. If it is analogous to that, it is not a direct effect on the pH itself, it is actually a channel effect rather than a pH neutralizing effect.

Peiris: If I recall, amantadine was tested in cell-based antiviral screening with SARS coronavirus. It had low activity at physiological levels. At the concentrations normally used to treat flu, it had no effect on SARS.

Skehel: Generally, amantadine just functions as a base if it is used at high concentrations. If it is blocking the M2 channel it is blocking the transport of protons.

Webster: Were antibodies used in early therapy of SARS?

Peiris: It was tried in Hong Kong, but there were no controlled trials so there was no way knowing whether it had a benefit or not in the small number of cases.

Osterhaus: We did experiments in monkeys with polyclonal immunoglobulin which had been purified, and this worked in a preventive way. Also, with a human monoclonal antibody that we used in the ferret model, we had good protection. Later on, in various experiments in mice and monkeys, it was shown that antibodies do have a protective effect when used preventively. In a clinical situation it may be different. In the monkeys with the human polyclonal antibody we showed a clear dose-dependent effect in terms of amount of virus still present.

Anderson: Treatment issues are important for SARS.

Skehel: Was the monoclonal against S?

Osterhaus: Yes. All the protective antibodies that have been tried in the animal models were neutralizing antibodies against the S protein. There is a whole panel of neutralizing human monoclonal antibodies against SARS coronavirus C, where the activities are known. But there is no interest at this time to have them mass-produced and stockpiled.

Webster: It is important to note that they are available.

Osterhaus: Yes, but the only problem is that in an outbreak we'd have to produce them.

Peiris: One problem here is whether there will be antibody cross-reactivity with the animal SARS-like coronavirus precursors in bats or even with the civet coronavirus.

Osterhaus: Panels have been made and work is ongoing to see how protective the effect is. They would only work in a preventive way, but in the absence of an approved vaccine this might still be important. In the animal models we showed that there is no such thing as antibody mediated enhancement. This is one of the things we were afraid of.

Kahn: I have a general question about the accessory proteins that you mentioned. There has been some attention on the accessory ORFs because this is where the deletion occurred between the viruses that were isolated from the civets and humans. What is known about this deletion? Apparently it created the two different ORFs. Are these ORFs responsible for the pathogenesis in humans?

Tan: There was a study on cell culture showing no difference in viral replication (Yount et al 2005). One study in the animal model also showed no difference (Wu et al 2005).

References

Duan J, Yan X Guo X et al 2005 A human SARS-CoV neutralizing antibody against epitope on S2 protein. Biochem Biophys Res Commun 333:186–193

Lu W, Zheng BJ, Xu K et al 2006 Severe acute respiratory syndrome-associated coronavirus 3a protein forms an ion channel and modulates virus release. Proc Natl Acad Sci USA 103:12540–12545

Oostra M, de Haan CA, de Groot RJ, Rottier PJ 2006 Glycosylation of the severe acute respiratory syndrome coronavirus triple-spanning membrane proteins 3a and m. J Virol 80:2326–2336

Wu D, Tu C, Xin C et al 2005 Civets are equally susceptible to experimental infection by two different severe acute respiratory syndrome coronavirus isolates. J Virol 79:2620–2625

Yount B, Roberts RS, Sims AC et al 2005 Severe acute respiratory syndrome coronavirus group-specific open reading frames encode nonessential functions for replication in cell cultures and mice. J Virol 79:14909–14922

Zhong X, Guo Z, Yang H et al 2006 Amino terminal of the SARS coronavirus protein 3a elicits strong, potentially protective humoral responses in infected patients. J Gen Virol 87:369–373

How the SARS experience has helped preparations for future outbreaks: the Taiwan experience, with emphasis on the successful control of institutional outbreak of influenza in 2003/2004 using a stockpile of antivirals

Ih-Jen Su

Division of Clinical Research, National Health Research Institutes, Tainan, Taiwan

Abstract. The experience and lessons learned from SARS in 2003 have driven Taiwan to prepare for the coming outbreak of SARS, pandemic influenza and other emerging infectious diseases. Several control measures were activated in the post-SARS period including central command and governance structure re-organization, improved scientific capability and laboratory diagnostics, surveillance and real-time reporting, and law revision and enforcement. Furthermore, the government implemented a policy to self-manufacture antivirals and vaccine for influenza and H5N1. The above measures proved to be effective for the control of dengue infection, seasonal influenza and enteroviruses in the post-SARS period. The measure most worthwhile to share with the world is the stockpile of 2.3 million dosages of the antiviral Tamiflu on November 2003 in the belief that the majority of influenza-like illness (ILI) should represent influenza in winter season and to avoid the confusion with SARS. A total of 68 ILI outbreaks were reported in 2003/2004 and Tamiflu was immediately given to the patients and their contacts. A total of 311 samples were obtained and 24% of them were laboratory-proved to be caused by influenza A. None of them were caused by SARS-CoV. All the ILI outbreaks were successfully controlled and no transmission was reported thereafter. In parallel to this control measure, the ILI prevalence and institutional mortalities due to ILI reduced remarkably in 2003/2004. Therefore, the control policy of ILI illness by antivirals proves to be effective in institutional outbreaks of influenza. This strategy can be considered for application in the case of H5N1 pandemic flu.

2008 Novel and re-emerging respiratory viral diseases. Wiley, Chichester (Novartis Foundation Symposium 290) p 89–98

The lessons we learned from SARS

Over-optimism (3 zero policy) and lack of support from WHO at the initial phase (before May 2003)

The SARS epidemic in Taiwan was marked by three distinct phases: the initial importation phase (before April 20, 2003), the second phase of explosive nosocomial outbreaks (April 21 to May 20, 2003), and the final containment phase (May 21 to July 5, 2003). The first imported SARS patient was a businessman who returned to Taiwan from China on February 25, 2003. The initial importation phase was controlled with success. All the SARS cases at this phase were from Guandong, Beijing, and Fu-Chien in China. However, the success of this phase drove the government to propose a propaganda of '3 zero—no case mortality, no secondary transmission, and no exportation' to mark the merit of SARS control, which later led to a delayed response to the index SARS case who had no history of travelling to China or Hong Kong. The second important lesson for Taiwan is the total lack of communication with WHO before April 2003, which led to the delay in revision of the WHO criteria for SARS, and even worse, the WHO criteria were used as a golden standard to exclude the index case even in the presence of repeated laboratory detection of SARS-CoV. Therefore, real time alert, and access to scientific information from WHO and other international reference centres are of critical importance to implement appropriate control measures at the initial phase. The importation of re-emerging pathogens and patients from China is always a big challenge for Taiwan.

Delayed reporting, inappropriate control measures, and poor public communication

In contrast to the success of the initial importation phase, the second phase of serial nosocomial outbreaks represented the startling and painful memory of SARS in Taiwan, when a large-scale nosocomial infection, 30 probable and 50 suspected SARS cases, were recognized in Taipei Municipal Hospital on April 21, 2003. The administrators in the hospitals were usually reluctant to accept the occurrence of SARS at the early stage and therefore slowed down their appropriate response, until the outside authority intervened. Understanding the attitude of the administrators drove the government to fine or punish the hospital administrators and, unexpectedly, this new policy led to an outburst of case reporting of SARS suspects from the hospitals, which subsequently shut down the laboratory and control system in early May. Furthermore, the poor communication among news media and a shortage of masks dampened the public attitude throughout the island. The improvement of public communication through regular television release of the real time status of SARS control and government policy starting from May 20 was effective in calming down the public panic.

Science contributes to control measures and wins the final battle

The dispatch of WHO officers to Taiwan on 4 May and the clarification of the incubation and transmission pattern of SARS on 6 May through the WHO telecommunication represent two major breakthrough in the control of SARS in Taiwan. Starting from 15 May, fever was recognized as an important indicator for SARS transmission and appropriate fever triage was implemented in each clinic and hospital throughout the island. A nation-wide fever screen was even implemented under the strong recommendation of the Nobel Prize Laureate, Dr Yen-Tzer Lee. The successful application of fever triage, patient isolation and trafficking measures then brought the SARS under control.

Border control and quarantine measures: should we do it again in influenza pandemic?

During the late SARS phase, infrared body temperature screening devices that can accurately measure body temperature when inbound and outbound travellers walk through at their usual pace were established at the two international airports in Taiwan. During the SARS outbreak, passengers with fevers were prohibited from boarding the aeroplane. A hospital near each airport was designated to house, diagnose and treat any passengers found with fever at the airport. Interestingly, in the summer through autumn of 2003, 15 cases of imported dengue and/or malaria were identified at the airports. In part, it is believed that such a fever screening system would serve to deter SARS patients from international travel, and could reduce the international spread of SARS. However, due to the difference in the timing of infectivity in relation to the time of symptom onset, how well this screening device can prevent international spread of infections other than SARS remains to be evaluated.

The policy to quarantine SARS contacts (level A) and those travelling back from the SARS epidemic regions (level B) led to the quarantine of a total of 130 000 individuals, leading to a tremendous economic loss. Retrospective analyses of the quarantine measures revealed that the criteria used for quarantine could be more specific through contact investigation. Since the transmission pattern of influenza is distinct from SARS, quarantine measures will not be effective for an influenza pandemic.

The post-SARS preparedness and policy

Strengthening the scientific capability

Twelve virology reference laboratories have been established throughout the island to keep a rapid and smooth chain of flow of clinical specimens from fever patients to the CDC central laboratory and reference laboratories beginning on August 19,

2003. 50 patients each week were selected from fever clinics to be tested for a variety of respiratory pathogens, including the SARS coronavirus, influenza, dengue and the Japanese encephalitis virus with the rapid antigen test. Currently, this laboratory surveillance system was kept for the monitoring of communicable diseases, syndromic reporting, and influenza and enteroviruses. Considering the importance of scientific capability in the identification of new pathogens in the outbreak of re-emerging infectious diseases, the government provided Taiwan CDC a total of 47 new positions for recruiting physicians and scientists to strengthen the scientific capability of case investigation and pathogen identification.

Strengthening alert, reporting, and the surveillance system

In general, fever patients are evaluated in designated fever clinics where laboratory specimens are collected and aetiological agents are identified with rapid tests as described above. The reporting of SARS patients was added to the pre-existing web-based reporting system along with all other reportable infectious diseases, and can be accessed by all regional hospitals and medical centres. In addition, a special telephone line has been installed for reporting any unusual infectious diseases, imported or otherwise, or in any unusual clusters by the general public or any practising physicians. Several pre-existing surveillance systems are undergoing evaluation and revision, including emergency room-based syndromic surveillance that would allow analysis for targeted signs and symptoms related to infectious diseases (e.g. fever, cough, respiratory distress).

Revision of law for infectious disease control

The SARS epidemic reactivated disease-control measures, such as the commanding and governance structure, the legal base for quarantine of contacts and compulsory isolation of SARS patients. The public health code was revised in March 2003 to place the implementation and enforcement of these control measures on a legal grounding. Further revision of the public health code has been undertaken to accommodate the complex issues concerning the protection of rights and freedom of individuals in the context of the overall well-being of the public.

Biosafety issues

In August and December 2003, two SARS cases were reported in Singapore and Taiwan, in two research virologists working on the SARS coronavirus. The Taiwan CDC invited international and local experts to evaluate and discuss biosafety issues regarding practice, training and regulation. National policy on monitoring and regulation of biosafety, as well as biosafety level III and IV practice standards is being formalized. A third incident of laboratory-acquired SARS that initiated transmission occurred in Beijing, and further stressed the importance of providing guidelines for biosafety standards and maintaining public health vigilance.

Stockpile of antivirals to control the institutional outbreaks of influenza

The overlapping of the symptoms/signs between SARS and influenza in the winter season constituted a big challenge for the control of the resurging SARS. In the belief that the vast majority of influenza-like illnesses (ILIs) in the winter season will actually represent influenza, Taiwan CDC formulated a flow chart for the proper management of febrile patients in the winter season 2003/2004. Those febrile patients were regarded as influenza and treated as this under appropriate isolation measure unless laboratory analysis proved the case as SARS. Considering the frequency of clustering outbreak of ILIs in the institutions which may raise unnecessary panic in the public, the government decided to stockpile antivirals (oseltamivir/Tamiflu) for a total of 2.3 millions dosages. In the winter of 2003/2004, the Taiwan CDC for the first time distributed an antiviral drug through public health channels to healthcare facilities for aggressive prophylactic therapy. Young children, elderly patients, and patients with a high risk for severe influenza infection should be evaluated early if a fever develops, and given antiviral drugs early when deemed necessary, especially the clustering in institutions. The goal is to reduce the demand for isolation hospital wards if a SARS resurgence occurs. In addition, all influenza A viruses isolated from potential SARS patients are sent to the Taiwan CDC for subtyping. In winter 2003/2004, a total of 68 institutional outbreaks occurred and rapid tests for influenza A and B were performed. After sampling, antivirals were immediately given to the patients and their close contacts. A total of 12000 dosages were given. 23% of the specimens were reported to be positive for influenza. None of those tested showed positivity for SARS-CoV. Interestingly, follow-up studies revealed a successful control of all the ILI outbreak and transmission was arrested in all 68 institutions. As a control, the ILI outbreak in two prisons from which no antivirals were given due to delayed reporting had successive transmission of ILI in their residents. Therefore, the application of antivirals to the early outbreak of ILI in institutions appears to be of great success. This measure can be considered for application in the case of an influenza pandemic.

Looking to the future

The public health system in Taiwan, as well as in many other countries, has geared up to minimize the adverse health impact of a possible influenza pandemic. It should be reiterated that prevention measures will have to be adjusted to account for the similarities and differences between SARS and influenza. Both SARS and pandemic influenza are likely to be zoonotic in nature, and establishing the capability for early detection in initial human cases is the key to preventing large-scale human transmission. While both diseases carry a considerable surge potential in terms of the number of patients and healthcare workers potentially affected, antiviral drugs that can be used for prophylaxis are available to fight influenza. However, the demand for antivirals may be high and will require countries to stock adequate supplies in advance. If transmission begins in human beings at any focal point, the speed at which influenza spreads will depend on how early it is detected, and how fast the international community can mobilize and deliver assistance, including providing antiviral drugs for prophylactic use. Therefore, in addition to a national preparedness plan, Taiwanese scientists and government officials are also actively seeking international collaborations with neighbouring countries in Asia. With the unexpected emergence of the H5N1 avian influenza in people during the winter of 2004, the preparedness plan in Taiwan for SARS has produced the additional benefit of consolidating the preparedness plans for possible influenza pandemics. Through the already existing viral laboratory network, methods for detecting the H5 influenza serotype among patients were quickly established in all the reference laboratories, and have become one of the routine diagnostic items for severe respiratory infection. The intensified efforts to identify aetiological agents for respiratory diseases have put these laboratories in a well-prepared state to detect the avian influenza virus when it occurs in human beings. In fact, early detection of any avian influenza virus in human beings is believed to be the key to influenza pandemic prevention and has been the main focus of concern for clinical laboratories in Taiwan.

Further reading

Center for Disease Control DOH, Executive Yuan, Taiwan 2004 CDC Health Advisory: severe acute respiratory syndrome (SARS) in Taiwan. 2004/2/1 (http://www.cdc.gov/ncidod/sars/taiwan17dec2003.htm) (accessed September 15, 2004)

Center for Disease Control, DOH, Executive Yuan, Taiwan 2004 Memoir of severe acute respiratory syndrome control in Taiwan. 2004/3/1 (http://www.cdc.gov.tw/sarsen) (accessed September 15, 2004)

Center for Disease Control, DOH, Executive Yuan, Taiwan 2004 Prevention and control of SARS in Taiwan. 2004/3/1 (http://www.cdc.gov.tw) (accessed September 15, 2004)

Center for Disease Control, DOH, Executive Yuan, Taiwan 2004 SARS website (http://www.cdc.gov.tw/sars/default.asp) (accessed September 15, 2004)

Chinese SMEC 2004 Molecular evolution of the SARS coronavirus during the course of the SARS epidemic in China. Science 303:1666–1669

Lee ML, Chen CJ, Su IJ et al 2003 Use of quarantine to prevent transmission of severe acute respiratory syndrome—Taiwan. MMWR Morb Mortal Wkly Rep 52:680–683

Ho M-S, Su I-J 2004 Preparing to prevent severe acute respiratory syndrome and other respiratory infections. Lancet Infect Dis 4:684–89
Wu H-S, Chiu S-C, Tseng T-C et al 2004 Serologic and molecular biologic methods for SARS-associated coronavirus infection, Taiwan. J Emerg Infect Dis 10:304–310
Yeh SH, Wang HY, Tsai CY et al 2004 Characterization of severe acute respiratory syndrome coronavirus genomes in Taiwan: molecular epidemiology and genome evolution. Proc Natl Acad Sci USA 101:2542–2547

DISCUSSION

Webster: One interesting issue is the use of steroids for treating SARS. There are people here who have experience of this. What are your impressions?

Su: My data are based on the high mortality from SARS, which is around 20%. Many of these patients die from complications such as bacterial and fungal infections. A high-dose course of steroids were routinely administered to the SARS patients in 2003. Many morbidities such as avascular bone necrosis were noted in SARS recoverers.

Webster: Should they have been used at all?

Su: I have reviewed autopsy samples from a patient in Thailand who died one month after H5N1 disease. In the autopsy samples, there is active virus replication in each alveolar epithelial cell, indicating that after one month the virus was still quite actively replicating. I reviewed the drugs that had been used on this patient, and found that the patient received steroids for two weeks until three days before death when antivirals were used. Perhaps the prolonged use of steroids was the reason for active viral replication in the autopsy sample.

Leo: In Singapore, we didn't use steroids a lot in treating SARS. Most of us are conservative in the use of steroids in acute infections. The steroid use became more common in the much later part of the SARS outbreak where we had the opportunity to do a few autopsies. Some of these patients died in the second to third week into their illness. We saw a lot of inflammation reactions with very few viral particles. We thought this could be due to over-exuberant inflammatory reactions that damaged the pulmonary tissues. So in the second half of the SARS epidemic we introduced steroids together with IVIG. We had about 15 cases who went through the treatment of using high dose steroid (methylprednisolone) together with IVIG. We saw a survival benefit but the number of patients was too small to give significance.

Peiris: In Hong Kong, there were two influences that promoted steroid use. First, the reports from clinicians in Guandong in February 2003 where they reported that the use of corticosteroids provided beneficial effect. The other is that some of the early cases of SARS in Hong Kong made a dramatic improvement after the use of steroids. This led to the increasing use of steroids in Hong Kong. Given the fact that this was a disease also effecting health care workers, there was the real pressure to do something to help the patients and this led to the use of steroids 'pre-emptively' early in the illness even before acute respiratory distress had set in. While it is clearly inappropriate to use corticosteroids in a pre-emptive manner, some clinicians have the impression that at least in some patients at certain stages of the diseases, there was clinical benefit. But there are no hard data backing this up. The other angle to consider is the question of ARDS, which is what these patients are dying of. The use of steroids in ARDS is still controversial. Some clinical trials suggest some benefit; others suggest none. But in summary, in the context of a viral disease where you don't have an antiviral agent of proven efficacy, using steroids is probably bad news.

Skehel: Did you screen for antiviral resistance in your trial?

Su: No evidence of drug resistance was discovered.

Webster: It would be a good opportunity with your use of antivirals to do this. What is the availability of the antiviral? Is it available over the counter (OTC)?

Su: The use of anti-virals needs prescription by the physicians or offered by CDC.

Webster: Some countries are turning to OTC delivery. In New Zealand they are making Tamiflu available over the counter. There is a lot of controversy about this.

Tambayh: You mentioned a big public health response. When you had a laboratory-acquired case, to what degree did you activate the public health response for a single case (Normile 2004)?

Su: That's an interesting question. There is a procedure that we follow. I received the notification at 10.30 pm and contacted CDC to discuss how to proceed, and at midnight we did the laboratory diagnosis simultaneously in two different laboratories. We also did a case investigation. At 4 am we got the laboratory information and they were SARS-positive. We convened a meeting with the CDC and experts to discuss how to prepare a response, how to handle the patient and how to handle the media communication. This was at 6 am. At 9 am we transferred the patient from the hospital to the SARS unit. Already the media noticed something unusual so they called us, and we arranged SARS team meetings at 10 am. We decided to inform Singapore, because this patient had just returned from a Singapore meeting. We reported to the WHO, Singapore, China and Japan at 10.30 am and had a press conference at 11 am. The same thing happened in May 2004 when the Beijing laboratory cases came out. We responded very efficiently to any notification to Taiwan CDC.

Lai: You also said that in the winter of 2003 after SARS, the cases of flu decreased. Is this a common experience outside Taiwan?

Su: We noted a steady decrease of ILI reported each year after 2003.

Tambayh: This was the year of the antigenically different flu strain that was not covered by the vaccine. Other countries saw an increase in flu cases that winter (Centers for Disease Control and Prevention 2004).

Lai: I don't know whether the reduction of flu in Taiwan in 2003 can be equated to the SARS control measures or not.

Webster: Didn't you see the same kind of thing in Hong Kong, with an upswing in apparent flu?

Peiris: In Hong Kong, after SARS was controlled in around June 2003, we saw a marked drop in the respiratory viral infection admissions to hospital. While aversion to hospital may contribute to this, hepatitis A admissions, for example, could serve as a control and there was no change in this. It seems that the social behaviour during SARS had impacted on respiratory viral transmission. In addition, after SARS, because there were more hospital outbreaks that were being investigated aggressively, more examples of hospital outbreaks due to flu, and also due to other viruses such as rhinovirus of parainfluenza virus, which hadn't previously been noticed came to our attention. This indicated how many of these respiratory viruses cause problems in hospitals.

Lai: If it is true that in all the countries that were affected by SARS the number of flu cases dropped whereas in other countries the flu cases increased, this might be telling us that the anti-SARS measures can be effective in controlling flu. This might be worth studying further.

Osterhaus: This may just be because of social distancing.

Peiris: Of course. People didn't go out and eat in restaurants, people wore masks, and there was an increase in hand hygiene. Even now, entrances to many Government offices and eating places in Hong Kong have alcohol rub dispensers at the entrance. There was a big change in social behaviour. We don't know which of these is most important. Is hand hygiene effective in reducing flu transmission? At the moment there are a number of studies across the world trying to address these questions.

Osterhaus: It's a large package of measures that apparently can be implemented in a crisis situation.

Webster: What steps were put into place in Taiwan?

Su: The government stockpiled 4 million N95 masks. There is a big warehouse with enough of them for an outbreak.

Webster: I also noticed that you put into place 25 BSL3s in Taiwan, which is a huge number.

Su: Yes, during the SARS period we had a big SARS budget for response. The maintenance of these laboratories has become a big issue after SARS: they cost about US$100 000 each year. Some of them have been converted to TB and HIV studies.

Peiris: Was there a large increase in the number of isolation beds in hospitals?

Su: Yes. About 500 negative pressure wards were created within three months.

Anderson: When you started your 25 BSL3 laboratories, did that include a training programme as well?

Su: Yes. We invited two experts from the US CDC and Japan to train our staff and to provide the guidelines.

Webster: You mentioned that there is a stockpile of H5N1 vaccine in place. Which clade was chosen for this?

Su: This was RG14 supplied from the UK. It is clade 1. The pharmaceutical division developed a drug to enhance viral production up to six times, so even in the cell culture the virus replicates pretty well.

Webster: Which cell culture do you use?

Su: MDCK cell line.

Webster: You also made an interesting comment that when it comes to the diagnosis of future strains of flu, Taiwan is two years ahead. Have you been correct in your selections two years ahead? What are they based on?

Su: We have a national surveillance system for ILI every year. It has been running for more than 10 years. On the basis of local surveillance data and comparisons with the WHO strain we came to this conclusion. The reassortment of influenza B virus usually first turns up in Taiwan.

References

Centers for Disease Control and Prevention (CDC) 2004 Assessment of the effectiveness of the 2003–04 influenza vaccine among children and adults–Colorado. MMWR Morb Mortal Wkly Rep 53:707–710

Normile D 2004 Infectious diseases. Second lab accident fuels fears about SARS. Science 303:26

General discussion I

Will SARS return, and why did it fail?

Webster: For this general discussion, I would like us to focus on whether or not SARS will come back. The reservoirs are still there. Is there still SARS in the civet cat, for example?

Peiris: When we look in markets, often we can find SARS-like coronaviruses in civet cats. But in the farms that supply the markets, they are largely negative. There have been few studies on wild-caught civet cats. The only one I know of was done in Hong Kong, looking at 35 of them. They are negative virologically and serologically. If we take this information together, it suggests that the civet cats are not the natural reservoir, but they may be the amplifier in this market setting. In 2005 two groups, in China and Hong Kong, found SARS-like coronaviruses in bats, but these quite distant to the human or civet coronavirus—perhaps too distant to be the immediate precursor. There may be a missing link somewhere.

Osterhaus: An intermediate species would be most likely. Too little work has been done on looking at the susceptibility of wild animals. Rodent species have not been looked into.

Webster: Considerable work was done on rodents in Hong Kong.

Anderson: The source of the virus from the outbreak remains uncertain. It could be bat SARS-like coronavirus that circulated in an intermediate species for sufficient time to acquire the changes in the outbreak strain. It could also be a more direct transmission from a bat with the parent to the outbreak strain/clade responsible for the SARS outbreak having not yet been encountered.

Peiris: That's fair.

Lal: How many different SARS-like coronaviruses are there?

Anderson: Six or seven have been identified so far in bats.

Osterhaus: Again, quite a bit of work has been done on wild animal species, but the most likely explanation is that you have a reservoir of this particular virus, which may be an intermediate species or a bat, and then the bringing together of these animals at the marketplace creates the possibility for large amplification. If the virus gets in there at a point in time, then it explodes. This is a concept that is not only true for the SARS coronavirus. The question is, why hasn't it happened before? There are good indications that due to the upsurge in economic wealth there has become an enormous market for these wild-caught rodents or even captive bred animals. This creates a whole niche of animal species that get into contact with each other.

Skehel: Is this the case in Hong Kong and China? Are these live markets novelties?

Peiris: Live poultry markets are widespread. But this type of market with a whole range of other mammalian species is quite common in southern China, but not in Hong Kong, at least in the recent past. They were common 20 or 30 years ago. Furthermore, the scale of operation of these markets has vastly increased, with wild-caught animals being shipped from many countries to service the demand and also some species such as civet cats being farmed. This is the issue. SARS must have arisen many times in past centuries, killed a few people and then burned itself out. But now when it arises the amplification factor is huge and the opportunity for human–human interaction is huge.

Osterhaus: Without the amplification it might not have spilled over to humans at all.

Anderson: When they did serological studies on workers in the marketplace, they found evidence of SARS antibody, and these people had no history of a SARS-like illness.

Holmes: I don't understand why you think there is a missing link. Looking at your data, the bat virus RP3 is very close to the SARS group. It is much closer than chimpanzee HIV is to human HIV, for example. That could be it. I would hesitate in thinking that there must be something else out there. You may have the right one, but it could be a strain that has evolved in itself.

Anderson: It is very close, but still sufficiently different to not be the direct parent strain.

Holmes: In my experience looking at many emerging viruses, that is as close as you would ever see for a donor–recipient pair. When you sample a reservoir population you will never get the absolute ancestor. What you are looking for is a common ancestor, and your estimation of the common ancestor times overlaps with the SARS epidemic.

Osterhaus: Another question is, how crucial is the civet? There will be another 10–20 species in which the virus has been.

Anderson: The data are good that a virus from bats was the source.

Osterhaus: This is trivial discussion. It doesn't matter if out there in these carnivorous species there are these viruses. If you find one species where the virus has been and then eradicate this species, then there are all the other species that are left.

Lai: The bat virus may not have been the immediate progenitor of SARS, but it does have the potential to become the source of another SARS outbreak.

Peiris: One of the practical implications is that one should consider banning the sale of live species in these markets.

Holmes: One of the basic rules of ecology is that the larger and more dense your population is, the more bugs you can carry, and the worse the bugs you can carry.

It is obvious that things with big population density—bats, rodents, some bird species—will carry more pathogens, and more virulent pathogens.

Osterhaus: You'd like to have these animals dead at the marketplace. The problem is once they are dead they can't be kept very long and good cooling facilities are needed.

Webster: I want to change the subject. Why did SARS fail?

Osterhaus: It was highly successful! Why do you think we are sitting here?

Su: During the post SARS period, we screened 3000 inhabitants in two communities, one in the north and one in the south of Taiwan. All turned out to be negative, even the community surrounding the Ho-Ping municipal hospital—the SARS hospital in 2003.

Webster: So SARS is a lousy transmitter.

Holmes: It was a decent virus. It was a successful emergent virus.

Kahn: Is there evidence that the virus became less virulent as the epidemic continued, and would one predict that the virus would mutate given the number of replication cycles that occurred during the epidemic? Was the mortality rate different at the end of the epidemic?

Lai: Maybe because of the peculiar transmission pattern of this virus, the public health measures to control it were particularly effective.

Kahn: But was the disease less severe at the end of the epidemic for the people who were infected?

Peiris: Implicit in your question is the concept that a successful pathogen becomes less virulent to its host.

Holmes: That isn't true: normally it is the other way round.

Lal: Isn't that the inherent nature of the plus-stranded viruses: they must keep on evolving. Naturally they would become less fit for the human host.

Osterhaus: It can go both ways. It can be an advantage to kill more people.

Holmes: The only rule is never to try to predict it.

Kahn: Would you predict there would be mutations?

Holmes: Yes. It's an RNA virus. It is temporal. We don't know what the selection pressures were, so it is impossible to predict. Virulence is a minefield. You could argue that high virulence aids transmission because there is more virus in the respiratory tract. You can argue any way you want.

Osterhaus: If you were to look over centuries, then there is an optimum for the host species and the virus to coexist, so eventually we would predict less virulence. But over a decade, this is not the case.

Holmes: However, the data don't support this classic idea that virulence goes down eventually. Looking at all the data together there is no overall trend.

Osterhaus: The SARS epidemic was too short for us to see a trend upwards or downwards, because we intervened and it disappeared.

Kahn: But we can't discount the possibility that there were some changes in the virus that occurred during the serial passage that may have triggered it.

Holmes: The one thing it will select for is transmission. It would never set any trait that reduces its transmission rate.

Osterhaus: That is overall transmission, taking into account that if a patient dies they don't transmit any more.

Lai: To judge whether SARS is a success or failure is premature: it may be hiding somewhere in nature. It may have the ability to hide somewhere in the body. Is there any evidence that it may have persisted, perhaps in macrophages, T cells or B cells? Murine coronavirus can persist in mouse brain for two to three years undetectable with no antibody response.

Peiris: We have no evidence of long-term persistence in humans, but other bat coronaviruses are persistent. There is one particular habitat that we sample on a regular basis. Each time we go there we find 10–15% of bats infected.

Holmes: You would need to do a mark/recapture study. Otherwise you could always argue there is reinfection taking place, rather than persistence.

Genetic and antigenic characterization of avian influenza A (H5N1) viruses isolated from humans in mainland China[1]

Yuelong Shu, Yu Lan, Leying Wen, Ye Zhang, Jie Dong, Xinsheng Zhao, Dayan Wang, Lihong Yao, Xiyan Li, Wei Wang, Xiuping Wang, Qi Wang, Shumin Duan, Jingjing Huang, Lei Yang, Hongjie Yu‡, Yuanji Guo, Weizhong Yang‡, Xiyan Xu†, Nancy J. Cox†, Xiaoping Dong, Yu Wang‡ and Dexin Li

Chinese National Influenza Center, National Institute for Viral Disease Control and Prevention, Chinese Center for Disease Control and Prevention, Beijing, P.R. China, †Influenza Division, National Center for Immunization and Respiratory Diseases, Centers for Disease Control and Prevention, Atlanta, GA, USA and ‡Chinese Center for Disease Control and Prevention, Beijing, P.R. China

Abstract. Twenty cases of H5N1 influenza virus infection in humans that resulted in 13 deaths were identified in China between November 2005 and July 2006. In this study, we performed genetic and antigenic analyses on the 16 H5N1 virus strains isolated from these infected persons. Sequence data obtained in our study indicates that isolates from southern and northern China each form unique sublineages among H5N1 viruses, with the isolates from southern China clustering in a novel clade. Antibodies raised against the H5N1 virus A/Vietnam/1194/2004, a candidate for current human pandemic vaccine production and clinical trials, had very limited cross-reactivity with the Chinese H5N1 isolates. Our study contains valuable information for global pandemic vaccine selection by providing evidence for the need to develop vaccine candidates selected from the H5N1 viruses isolated from these more recent human cases from China.

2008 Novel and re-emerging respiratory viral diseases. Wiley, Chichester (Novartis Foundation Symposium 290) p 103–112

Influenza, a type of respiratory disease, has inflicted human beings with severe disasters and caused worldwide pandemics periodically. During the 20th century, four large mutations occurred to influenza viruses, resulting in four worldwide pandemics, i.e. the H1N1 (Spanish influenza) in 1918–19, H2N2 (Asia

[1]This paper was presented at the symposium by Li Xin of the Chinese National Influenza Center, National Institute for Viral Disease Control and Prevention, Chinese Center for Disease Control and Prevention, Beijing, China.

influenza) in 1957, H3N2 (Hong Kong influenza) in 1968, and reappearing A1 H1N1 sub-type (Russian influenza) in 1977. China is publicly recognized by the world as a place frequently threatened by influenza. In 1997, an avian influenza pandemic caused by H5N1 broke out in Hong Kong, resulting in 7000 chickens killed and 18 people infected with 6 deaths. Up to May 31st, 2007, influenza A H5N1 virus has invaded 12 countries with 310 confirmed human cases and 189 deaths (the fatality ratio greater than 50%) (World Health Organization 2007).

In China, 24 human cases with 15 deaths have been confirmed through the Chinese avian influenza (AI) surveillance network since November 2005, with the first human H5N1 infection being reported in November 2005 (Yu et al 2006). The confirmed cases are mostly centralized in nine provinces in south China, while two cases are located in north China. Since the outbreak of avian influenza (H5N1) in animals and the confirmation of the first human case, an AI (H5N1) surveillance network has been built up based on the seasonal influenza surveillance system, including 63 network laboratories and 197 hospitals at a national level, and the number of laboratories will be increased by 20 in the following year. In order to control the quality of laboratorial detection in network laboratories, a quality control system was developed, providing necessary reagents and equipment, reference reagents, etc. Moreover, periodical supervision and technical training is performed once a year. With the help and support of local CDCs, specimens were collected and transported to CNIC, and serological study and phylogenetic analysis were performed.

Antigenic analysis

Serological study is a typical method for AI laboratorial diagnosis, mainly including a haemaglutinin inhibition (HI) and micro-neutralization (MN) assay. The HI assay is considered the gold assay for influenza virus antibody detection. The haemagglutinin (HA) protein on the surface of influenza virus contains a structure that can recognize and combine receptors on the host cells. Glycoprotein and lipoprotein with sialic acid, as the content of receptors of influenza virus, appear on the surface of red cells in many mammals and birds, therefore, they all can be recognized and combined by influenza viruses, resulting in a haemagglutination phenomenon.

The MN assay, a highly sensitive, highly specific serological method, is used to test specific antibody titre level in serum. Quantitative virus is mixed with multiple proportional sera dilution. Then, virus–antibody mix is incubated in sensitive a host (tissue cells, chicken eggs or animals). The MN assay is based on virus infectivity, virus specific antibody binding with proteins on surface of viruses, therefore, the viruses losing the infectivity to host cells. Briefly, the MN assay contains three steps, virus titration, virus neutralization and ELISA.

TABLE 1 Antigenic analysis of human H5N1 isolates by HI assay using ferret serum

	Reference ferret serum				
	VN/1203	*IND/5*	*TK/15*	*AH/1*	*AH/22*
A/Vietnam/1203/2004	**320**	20	10	40	40
A/Indonesia/5/2005	20	**640**	40	160	80
A/Turkey/15/2005	80	320	**1280**	80	20
A/Anhui/1/2005	40	320	10	**1280**	640
A/Anhui/2/2005	20	80	5	640	**320**
A/Guangxi/1/2005	20	80	5	640	320
A/Fujian/1/2005	10	80	5	160	160
A/Sichuan/1/2006	40	160	10	1280	640
A/Jiangxi/1/2005	80	640	40	2560	1280

The antigenic character of human H5N1 isolates from China has been compared with that from Vietnam, Indonesia and Turkey by HI assay using reference ferret serum raised against representative H5N1 viruses. Ferret antisera were kindly provided by the Influenza Division, Centers for Disease Control and Prevention, Atlanta, Georgia, USA. Survivor sera of confirmed human H5N1 infection from Anhui, Guangxi, Fujian, Sichuan and Jiangxi provinces, respectively, were collected 4 weeks after the onset of illness. As shown in Table 1, the H5N1 human isolates from China were well inhibited by ferret antisera raised against Anhui/1/2005 and Anhui/2/2005 viruses isolated from southern China, but were not well inhibited by antisera raised against H5N1 viruses isolated from Turkey, Vietnam and Indonesia, with at least fourfold reductions occurring in levels of HI titres compared to the levels of homologous titres. These results indicated that Chinese human H5N1 isolates are antigenically closely related to each other, but they are antigenically distinguished from H5N1 isolated from Vietnam, Indonesia and Turkey. Additional HI analysis using chicken antisera raised against representative H5N1 viruses also revealed antigenic differences between the two groups (Table 1).

Microneutralization assays using serum samples collected from confirmed human cases were consistent with the results obtained from the HI tests (Table 2). These findings demonstrate the considerable antigenic divergence of HPAI H5N1 viruses.

Genetic analysis

Influenza virus' RNA genome has 8 segments and each segment encodes 1–2 proteins. Since its gene is composed of several segments, the virus' RNA polymerase does not have the function of error correction. Therefore, it is easy for mutation to happen in virus genes. The spontaneous point mutation in genome

TABLE 2 Antigenic analysis of human H5N1 isolates by MN assay using confirmed human case serum

	Confirmed cases	
	*case1**	*case2#*
A/Anhui/1/2005	**160**	40
A/Xinjiang/1/2006	40	**320**

*Case 1 is a surviving confirmed human H5N1 case from Anhui province (southern China), the serum was collected 4 weeks after the illness onset.
#Case 2 is a surviving confirmed human H5N1 case from Liaoning province (northern China), the serum was collected 4 weeks after the illness onset.

often causes antigenic drift. Thus, the infection of cells by two virus strains of different subtypes can rearrange the genome. This can result in antigenic shift and then appearance of new subtypes. The result of comparison of the genetic sequence of 1918 H1N1 virus and H5N1 virus concluded that H5N1 avian influenza virus may trigger an influenza pandemic through one or several steps of variation.

In this study, respiratory specimens were collected from patients on days 5–11 after onset of illness. Samples included gargle, nasal and/or throat swabs, tracheal aspirates and phlegm, saliva, as well as lung tissues from autopsy collected from different individuals following a standard protocol. The viruses were isolated and propagated in the amniotic and/or allantoic cavities of specific pathogen free (SPF) embryonated chicken eggs as described under biosafety level (BSL) 3 enhanced containment in accordance with guidelines from the National Institutes of Health and the Centers for Disease Control and Prevention (Interim CDC-NIH Recommendations, *http://www/cdc/gov/fluh2n2bsl3.htm*). Sixteen H5N1 virus isolates were obtained from 13 fatal and three surviving human cases.

Viral RNA was extracted using the RNeasy Mini Kit according to the manufacturer's protocol (Qiagen, Valencia, CA). cDNA synthesis and PCR amplification of the coding region of the eight gene segments were carried out using the one-step RT-PCR kit (Qiagen, Valencia, CA) with gene specific primers (primer sequences available on request). The PCR products were then purified using the QIAquick Gel Extraction Kit (Qiagen, Valencia, CA) and used as templates for nucleotide sequencing. Sequencing reactions were performed using the ABI BigDye Terminator Sequencing Kit with reaction products resolved on a MegaBACE1000 DNA sequencer (Applied Biosystems Foster City, CA). Nucleotide sequences were analysed using DNASTAR (Lasergene, Madison, WI). The phylogenetic tree was generated by the neighbour-joining method using the Mega3.1 program.

Phylogenetic analysis revealed that the HA gene of the H5N1 viruses diverged into several distinct clades (Fig 1). Clade 1 contains viruses isolated from Vietnam,

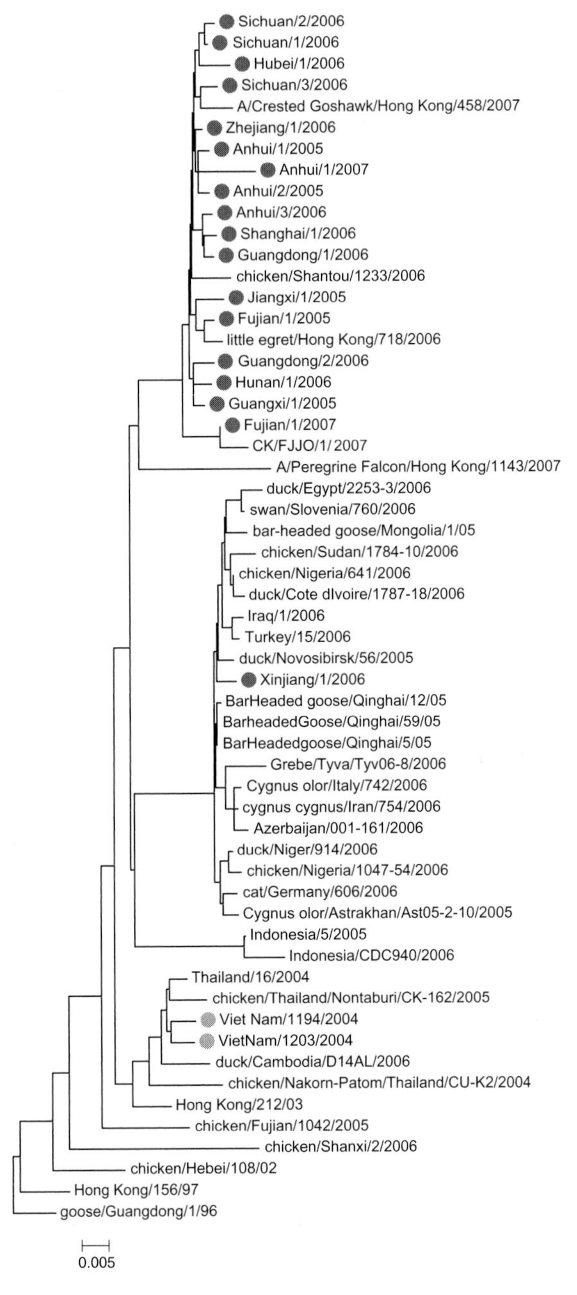

FIG. 1. Phylogenetic tree generated by the neighbour-joining method using the Mega3.1 program and the bootstrap value 1000. The relationships of HA genes of representative influenza A H5N1 viruses from different regions are shown. The dark grey dots denote viruses isolated from humans in China; lighter grey dots denote viruses isolated from Vietnam and that were recommended as vaccine strains by the WHO.

such as the virus A/Vietnam/1194/2004. Clade 2 viruses segregated into three sub-clades that are represented by A/Bar-head goose/Qinghai/5/2005, named clade 1.1, Indonesia/5/2005, named clade 1.2, and Anhui/1/2005, named clade 1.3. The single human H5N1 virus from north China, A/Xinjiang/1/2006, belongs to clade 2.1 and the isolates from southern China clustered together with Anhui/1/2005 (Chen et al 2004, 2006), belongs to clade 2.3.

The other seven segments (including NA, PB2, PB1, PA, NP, NS and M) were all sequenced and phylogenetically analysed. A clear reassortment was observed in Jiangxi isolate, A/JX/1/2005, that is reassorted by five strains, A/VN/1194/2004, A/BHG/5/2005, A/IND/5/2005, A/AH/1/2005, and A/CK/GXNN/6/2004. In addition, according to the sequence analysis of HA and NA genes, no popular amino acid mutation was observed that associated with resistance to amantadine, rimantadine, oseltamivir or tamiflu, etc., in their M2 (at position 25–42) or NA (at position 274) proteins although most patients received up to two weeks of treatment with amantidine, rimantadine or oseltamivir (at a daily dose of 40–150 mg). One exception of a point mutation in the M2 gene at position 31 (S31N) was observed in a recent isolate from Fujian province.

References

Chen H, Deng G, Li Z et al 2004 The evolution of H5N1 influenza viruses in ducks in southern China. Proc Natl Acad Sci USA 101:10452–10457
Chen H, Smith GJD, Li KS et al 2006 Establishment of multiple sublineages of H5N1 influenza virus in Asia: implications for pandemic control. Proc Natl Acad Sci USA 103:2845–2850
World Health Organization 2007 Cumulative number of confirmed human cases of avian influenza A (H5N1); available at *http://www.who.int/csr/disease/avian_influenza/country*
Yu H, Shu Y, Hu S et al 2006 The first confirmed human case of avian influenza A (H5N1) in Mainland China. Lancet 367:84

DISCUSSION

Webster: The fact that the Shanxi variant of H5N1 from northern China is significantly different from the southern strains that you would need a different vaccine is probably a concern globally.

Xin: Yes, from the serological study and gene analysis this strain is quite different from the strains isolated from southern China.

Osterhaus: We need to define what the criteria are for having vaccine prototype strains made. Are you going to cover all the clades that are circulating?

Webster: That's an important question. At this stage the philosophy is to cover the major clades that are circulating. Is this another dominant clade that is emerging? That's what I was getting at with my question. I'm not quite sure from the data that it is.

Osterhaus: If we are going to get a pandemic outbreak of H5, the key question is which strain it will be. Whatever we have in terms of different strains that are sufficiently different, we have to cover them all.

Webster: That is the strategy that is being followed by the WHO at this time.

Osterhaus: The question immediately arises: are we going to go on forever doing this? Why aren't we doing it for H9 and possibly others?

Webster: That is a good question. H5 is the virus that is most threatening and which has the most human cases, with over 50% mortality.

Osterhaus: If you were to have proper adjuvant systems, the question then arises, are we continuing to focus on every new subclade that will arise, or are we going to take a broader view?

Webster: That's what we have been doing for human influenza for many years. Until we have a proven, broad-based vaccine that works in humans, we have no other option.

Smith: You need a seed strain ready for each of these. Even if there is a great adjuvant, you might as well use the vaccine against the strain that is circulating.

Skehel: How confident are you that there are only 24 cases of H5 in China?

Xin: The surveillance system in China is good, so we are reasonably confident. At the moment we are doing the serological studies for people in contact with poultry. It is a cohort study of a representative sample. This is in collaboration with the surveillance network.

Peiris: You mentioned that you had 197 surveillance hospitals. How many of these cases came from those surveillance hospitals, and how many from others?

Xin: I haven't calculated the percentages. Most of the cases come from the countryside, and people are treated as if they have normal influenza in the countryside hospital, and then they may go to the hospitals in the city.

Anderson: Do you test all the patients with serious hospitalized pneumonia?

Xin: Yes. Suspected influenza cases are tested, and so far we have tested around 600 specimens.

Osterhaus: Is the suspicion only for people who have been in contact with birds, or anyone with severe pneumonia?

Xin: These 600 specimens are those who have been reported from the hospital. Surveillance in the countryside for those in contact with poultry is another study.

Leo: Have all your H5N1 cases had definite contact with poultry?

Xin: No.

Webster: One of the questions that is not resolved is the absence of serological positivity for family members with an H5N1 member. Is serology being done in all these families?

Xin: Yes, for close contacters and also healthcare workers. We don't have the results yet.

Webster: So the only serological positives are those that have recovered from H5N1 infection?

Xin: Yes.

Webster: What is the treatment if you have a diagnosis of H5N1 influenza?

Xin: We use antivirals. Which one depends on the doctor. We have studied the drug resistance for the antiviral drugs and there is none yet, not even for amantadine.

Kahn: What is the time between when the sample is obtained and the diagnosis is made?

Xin: It is immediate. Reporting has to be done in 24 h.

Osterhaus: What is the likelihood of identifying people who have been infected by serology?

Peiris: In confirmed cases, as long as you have a convalescent serum after two weeks, they are invariably positive. I am not really sure about the horse erythrocyte test, but from what I understand this seems to correlate well with the neutralization test. In population surveys of close contacts of patients or in people who have been working with poultry, the overall H5 sero-positivity has been very low.

Osterhaus: How do we interpret that? We had a similar situation with the H7 outbreak. A lot of people were exposed, and the HI was insensitive. We found quite a number of positives, but in addition to this quite a number of people who we thought should have antibodies were shown to have low titres.

Peiris: It is difficult to interpret this. Usually we set the cut-off at 40–80.

Osterhaus: The problem is, if you then look at the general population who have not been in contact with poultry we don't see the low level titres that are seen exclusively in the poultry workers. Is this exposure to antigen?

Peiris: In the Hong Kong population, we still see these low titres in unexposed people. In over-65s some have pretty strong titres to H5 without exposure. There is a cross-reaction problem coupled with a poor response problem.

Osterhaus: Is it cross-reaction? Ken Shortridge has some data where he claims this was exposure.

Wood: This was using SRH data. This test is known to be cross-reactive.

Webster: I see where you are going with the sensitivity of the assay.

Osterhaus: From the risk evaluation this is important. In the Asian data in general I suspect we are not seeing the tip of the iceberg. If we only have a couple of hundred of people who come down with the disease, and we don't see it move more broadly, this means that the whole discussion about this virus having been around for 10 years and there hasn't been a pandemic should be seen in another light, because only a couple of hundred people have been infected.

Webster: So we don't have a satisfactory assay for detecting the virus in populations.

Peiris: It may be correct that this assay is not sensitive enough, we can't rule out the other possibility, which is that the vast majority of people are not getting infected.

Osterhaus: I agree. We don't have enough data yet. One of the difficult points there is that our serological assays may not be good enough.

Wood: Even if you look at post-infection ferret sera, there are low levels of antibody. Perhaps it is not a very immunogenic infection.

Hoffmann: It depends how you define the immune response. Current assays are based on the detection of antibodies in serum. I think we need more information on whether the failure to detect antibodies is because of limitation of the assays or indeed low immunogenecity of some antigens.

Anderson: There are good data that most of the people who have a confirmed infection have a good, clear antibody response. I am not aware of any viral infections where we see a substantial difference in antibody response related to the severity of the illness. People get infected and get an antibody response. There is a distribution but it is not related to the severity of illness. If you get over the threshold needed to induce an immune response, it seems the responses are similar. The poultry workers may be exposed to a variety of avian influenza viruses, and there could be some level of cross-reactivity. But I'd wonder whether those with low antibody titres were not actually infected with the virus you are testing for.

Osterhaus: In the Netherlands it is a different story. The poultry workers were just people who were rounded up to do the culling, so they had not been exposed to chickens before.

Anderson: They may not have had a live virus infection if they did not have a good antibody response.

Su: In our model of EBV-associated haemophagocytic syndrome, there is a strong CTL immune response. However, the antiviral B cell response is quite weak and delayed. Only about 30% are positive for anti-EBV antibodies. In the H5N1 infection there is also an overt CTL response but a weak anti-viral response.

Osterhaus: I would be surprised. EBV is a chronic infection. We are talking about an acute infection with a lot of antigen being produced. I don't think there is any other acute respiratory virus infection where you wouldn't get a proper antibody response, even if you get a good CTL response.

Peiris: In general, are we sure that it is possible to have aysmptomatic infection without generating a detectable antibody response, for example for measles?

Osterhaus: It doesn't happen. You can't have measles infection without an antibody response.

Anderson: But if you don't have a way to detect infection, how can you tell?

Wood: I know that human H5N1 vaccines are being developed in China. What stage are they at, and how will they be used? Also, what H5N1 vaccines are being used in poultry?

Xin: It is true that China has developed its own vaccine. This is using an Anhui strain from south China. One company is developing a vaccine using a Vietnam representative strain and have finished phase I trials.

Osterhaus: Is that an adjuvant vaccine?

Xin: Yes, but I don't know which one.

Wood: The company based in Beijing, Sinovac, uses alum. Is there a plan to stockpile vaccine?

Xin: Yes. I don't know what the plans are for vaccinating poultry, but I know that one has been developed and is being used.

Webster: Poultry vaccine is widely used in China. The available knowledge is that it is the reverse genetics based H5N1 vaccine that is being used, and the quality is high. This is the same vaccine that has been used successfully in Vietnam on poultry.

Osterhaus: The poultry vaccines are not like human vaccines. They ones used for poultry give a broad protection and are not purified. The old strains that are being used by some of the commercial producers give very good cross protection against the current strains.

Wood: They use adjuvants that wouldn't be tolerated in humans.

Webster: But there is more than one company making vaccines. Some are of suspect quality. The problem is universal: we don't have standardization for antigen content of poultry vaccines globally.

Osterhaus: It is difficult to do an antigen standardization, because the adjuvants are a major part of the vaccine. The standardization of poultry vaccines is not being done on the antigen content, but on the immune response it elicits in the target species.

Webster: The indications are from the studies in Vietnam that the poultry vaccines are wonderful in chickens, but failed in some strains of ducks.

Emerging infectious diseases and the animal–human interface

J. S. M. Peiris*†‡ and Y. Guan*†

**The Department of Microbiology, The University of Hong Kong, Hong Kong SAR, †The State Key Laboratory of Emerging Infectious Diseases, China and ‡HKU-Pasteur Research Centre, Hong Kong SAR*

Abstract. Major emerging infectious diseases of humans continue to arise from animals, a fact well illustrated by SARS and avian influenza H5N1. Changes associated with our globalized lifestyle facilitate such zoonotic transmission. In order to be better prepared to predict and prevent future emerging diseases, we need to better define the viral flora in domestic livestock and wildlife, and better understand the biological and ecological determinants that allow or limit inter-species transmission of microbes. Pandemic influenza and SARS (and related coronaviruses) are likely to prove to be productive case-studies in this regard. Confronting emerging infectious threats requires a multi-disciplinary response, spanning the sectors of human and animal health, wildlife and environment, and the combined resources of government agencies and academics. It requires specialized expertise in the relevant fields but also requires those who can bridge interdisciplinary and organizational divides. It is important that such inter-disciplinary research is nurtured and facilitated.

2008 Novel and re-emerging respiratory viral diseases. Wiley, Chichester (Novartis Foundation Symposium 290) p 113–127

Over the last century, the conquest of infectious disease played an important role in improving the quality of life and life-expectancy worldwide. While these advances lulled many into a false sense of security, Rene Dubos was prescient when he stated that 'human destiny is bound to remain a gamble, because at some unpredictable time and some unforeseeable manner, nature will strike back'. While this applies to many aspects of the human condition, it is certainly also true of infectious disease. Scourges of the past have been controlled or even eradicated in some instances, but changes associated with our globalized and interconnected lifestyle have provided microbes the opportunity to strike back. Over two thirds of recently emerged infectious diseases are zoonotic in origin: the emergence of bovine spongiform encephalopathy (BSE) in the UK, the introduction of West Nile virus to the USA and the emergence of Nipah virus encephalitis in Malaysia and more recently in Bangladesh (Kuiken et al 2005). Perhaps the most dramatic examples are the emergences of SARS and H5N1 'bird flu'. Arguably, none of these outbreaks could have occurred 50 years ago. Even if they had, they certainly would not have had the same global impact in terms of the rapidity of spread and public anxiety.

In November 2002, SARS emerged as an 'atypical pneumonia' in Guangdong Province, China. It caused continuing outbreaks of disease, with clusters of cases occurring within families or in health care facilities in Guangdong. One infected individual travelled to Hong Kong and stayed in a hotel there for one just day during which time he transmitted the infection to 17 others. Many of these then travelled on to initiate outbreaks of infection elsewhere within Hong Kong and in Vietnam, Singapore and Toronto. Within weeks the disease had spread to affect over 8000 people in 29 countries across five continents with an overall case fatality of 9.6%. This was perhaps the most dramatic illustration to date of the rapidity with which infectious diseases can spread. Unusually, 21% of all cases occurred in health care workers and health care facilities were a major hub for the amplification and dissemination of disease. The disease impacted not only on human health but also on the economy and the collective psyche of the inhabitants in affected regions (Peiris et al 2004).

SARS was caused by a novel coronavirus (SARS CoV) not previously endemic in the human population (Peiris et al 2004). Coronaviruses are single stranded RNA viruses of positive sense. Bats appear to be the natural reservoir of the precursor of the SARS CoV while small mammals such as civet cats, raccoon dogs, Chinese ferret badgers and similar small mammals held for sale within live wild-game animal markets in Guangdong were the amplifiers of infection (Guan et al 2003, Lau et al 2005). These markets were the interface for the initial zoonotic transmission events that over time allowed the virus to adapt to efficient transmission within humans (Xu et al 2004). These markets offer many diverse and exotic species for a burgeoning demand for exotic foods by an ever-more affluent population in one of the most rapidly developing regions of the world. An extensive network of international, often illegal trade and the 'farming' of game animals has developed to service this demand (Bell et al 2004). The markets and associated trade provide the ideal milieu for inter-species transmission of animal viruses and for their zoonotic transmission to humans. One aspect of the adaptation of SARS to efficient human-to-human transmission was adaptation of the viral surface spike protein of SARS CoV to bind more efficiently to the human ACE-2 molecule, the key functional receptor for this virus (Li et al 2006). While the human-adapted SARS CoV isolated during the later phase of the outbreak efficiently binds to human ACE-2, the civet-virus binds poorly and the precursor virus found in bats is unlikely to bind human ACE-2 at all. Whether these viruses use ACE-2 or another receptor in the reservoir animal host is unclear.

Just as dramatic as the emergence of SARS was its control by the simple public health measures of case detection and isolation. This was due to a fortuitous characteristic of the infection in which virus transmission was relatively less common in the first five days of illness (Lipsitch et al 2003). This in turn appears to be because viral load in the upper respiratory tract is low in the early stages of illness and does not peak until about day 10 of disease (Peiris et al 2003). This feature of SARS allowed the application of public health measures to interrupt transmission within the community. This is unlikely to occur in other viral infections such as pandemic influenza where transmission is known to occur early in the disease (Fraser et al 2004).

While only two human coronaviruses, 229E and OC43 were recognized prior to the emergence of SARS, the attention focused on this group of viruses following the emergence of SARS led to the discovery of two more human coronaviruses, NL-63 and HKU-1, each contributing significantly to disease burden (van der Hoek et al 2004, Woo et al 2005). Similarly, the hunt for the animal reservoir of SARS led to the discovery of other novel coronaviruses in bats, in addition to the precursor of SARS CoV (Lau et al 2005, Poon et al 2005). Indeed, phylogenetic analysis suggests that coronaviruses in bats are in 'evolutionary stasis' and that, in addition to SARS CoV, many mammalian group 1 and group 2 coronaviruses also originated from bat coronaviruses (Vijaykrishna et al 2007). Thus bat coronaviruses have not only provided the precursors for the emergence of SARS CoV, but appear to have been the precursors for many human and other mammalian coronaviruses as well.

Influenza A viruses are also RNA viruses, but with an eight-segmented negative sense single-stranded genome. Aquatic waterfowl are the reservoir of the widest diversity of influenza subtypes (haemagglutinin subtypes H1–16 and neuraminidase subtypes N1–9). A limited number of influenza subtypes have become established in mammalian species including humans (H1N1, H2N2, H3N2 and H1N2), pigs (H1N1, H3N2) and horses (H7N7) (Webster et al 1992). Influenza pandemics arise when an avian virus directly adapts either through mutation or by reassortment of its segmented genome with a pre-existing human influenza virus and gives rise to a virus that can sustainably be transmitted from human-to-human. Southern China appears to the epicentre for pandemic emergence (Shortridge & Stuartharris 1982). In addition to pandemics, other animal influenza viruses may occasionally transmit zoonotically to humans including the subtypes H5N1, H9N2, H7N7, H7N3 and others (Peiris et al 2007).

Like SARS, the emergence of the recent highly pathogenic avian influenza (HPAI) virus H5N1 also was from Guangdong. The first known virus of this H5-sublineage was detected in disease outbreaks in geese in Guangdong in 1996 (A/Goose/Guangdong/1/96) (Xu et al 1999). Reassortants of this virus caused outbreaks in chickens in Hong Kong in 1997 (Guan et al 1999) and were zoonotically transmitted to humans, killing 6 of the 18 persons affected by it. That outbreak was aborted by the slaughter of all live poultry in Hong Kong in December 1997. However, the precursor A/Goose/Guangdong/1/96-like virus continued to circulate in geese in southern China. This virus again reassorted with influenza viruses from aquatic avian species giving rise to a range of reassortants, some of which readapted to become endemic in chickens (Li et al 2004). H5N1 influenza reassortants are designated as genotypes on the basis of the constellation of the 8 viral genes. One of these genotypes (Z) became dominant in poultry across the Asian region and gave rise to the panzootic that occurred in south-east Asia from 2003 onwards. This virus has now become entrenched within the poultry of many countries. In 2005, bird migration was likely the vehicle for the spread of the H5N1 virus to the middle-east and to Europe (Chen et al 2005, Ducatez et al 2006). However, movements of poultry and poultry products remain the main mechanism for its endemicity in poultry flocks in many other Asian countries (Smith et al 2006). There is now good evidence that live-poultry markets, which are widespread in many Asian countries amplify, maintain and disseminate such viruses and thereby contribute to its endemicity (Kung et al 2003, 2007). The widespread practice of keeping 'backyard-poultry', the occurrence of asymptomatic infections in ducks (which serve to disseminate infection silently), taken together with the lack of credible and reasonable compensation for affected poultry farmers makes it difficult to eradicate H5N1 from flocks once it has become entrenched in them (Hulse-Post et al 2005). This illustrates the importance of a multi-dimensional approach, rather than focusing exclusively on just one of several strategies (e.g. vaccination of poultry, detection and depopulation) for controlling H5N1 in endemic situations. In particular, it is important to identify the key drivers of virus maintenance and dissemination and apply strategic and evidence-based interventions to interrupt transmission. A good example of the success of such an integrated strategy are the control measures applied to the control of avian influenza in Hong Kong where live poultry markets were shown to amplify, maintain and disseminate virus infection (Sims et al 2003). The parallel with SARS in relation to the role of live-animal markets in the emergence and maintenance of H5N1 is striking.

Those humans diagnosed with H5N1 disease appear to develop an unusually severe pathology causing a fulminant viral pneumonia and acute respiratory distress syndrome. The virus can clearly replicate efficiently in the respiratory tract and also in other organs, including the gastro-intestinal tract (Beigel et al 2005, Gu et al 2007). However, given the widespread endemicity of H5N1 virus in poultry flocks across the region, it is clear that exposure alone does not explain the observed epidemiology of human H5N1 disease (Vong et al 2006). So far, the H5N1 virus has been extremely inefficient in infecting humans, and the vast majority of those who are heavily exposed to the virus via sick poultry do not get ill or even seroconvert to the virus (Vong et al 2006). On the other hand, a small proportion of those who get H5N1 disease have little or no exposure to widespread poultry deaths (Mounts et al 1999). Exposure is therefore a necessary but not sufficient cause for human H5N1 disease (Peiris et al 2007) and there are other host barriers or susceptibility factors at play in humans. Genetic susceptibility of the few or unexplained mechanisms of resistance preventing infection of the many are plausible hypotheses that remain to be systematically tested.

As with SARS, predilections for specific receptor binding are also believed to play a role in the avian–human species barrier of influenza viruses (Ito & Kawaoka 2000). Avian influenza viruses preferentially infect cells carrying sialic acid with an α2'3 linkage found in avian cells rather that the sialic acid with an α2'6 linkage that was believed to be dominant in human respiratory epithelium. This raises the question of how the avian H5N1 virus is able to infect humans in the first place? Recently however, it has been reported that the epithelium of the human lower respiratory tract bears the 'avian' α2'3-linked sialic acid, although the upper respiratory tract does not (Shinya et al 2006, van Riel et al 2006). This suggested the hypothesis that H5N1 was unable to gain a foothold within the human upper respiratory tract and had to reach the lower respiratory tract in order to initiate virus infection. However, *ex vivo* cultures of nasopharyngeal epithelium are efficiently infected by the avian H5N1 virus in the absence of the avian α2'3 receptor (Nicholls et al 2007), and there is also autopsy evidence that H5N1 virus can also infect the tracheal epithelium which appears to lack the α2'3 receptors (Gu et al 2007). These observations suggest that the current receptor paradigm needs re-evaluation. Furthermore, some human H5N1 virus strains have acquired capacity for binding α2'6-linked sialic acids but still fail to transmit efficiently in humans (Yamada et al 2006). A mutation in the PB2 gene (Glu627Lys) is associated with more efficient replication in the upper respiratory tract (Hatta et al 2007). However, a complete understanding of the critical combination of viral factors that determine human transmission still appears to elude us.

Surveillance at the animal–human interface

Most emerging infections diseases are zoonotic in origin and it is important to have a better understanding of the ecology of animal viruses and better surveillance at the animal–human interface (Kuiken et al 2005). The majority of available data on animal viruses is related to those causing economically important diseases in livestock. However, many emerging human pathogens do not cause overt disease in the livestock hosts, and in some cases these viruses have emerged from wildlife rather than from domestic livestock. Reliance on surveillance information from veterinary sources, which typically relates to diseases in domestic livestock, while necessary and very important, may not be sufficient to provide warning of future emerging disease threats. We need surveillance of healthy as well as diseased animals and from wildlife as well as domestic livestock, and this poses a formidable challenge. We clearly need closer collaboration between individuals and institutions with an interest in human health, animal health and wildlife. Technological developments such as the use of pan-virus DNA arrays and recent advances of high-throughput sequencing methodologies provide tools that can be applied to explore the microbial ecology of the animal world in ways that do not depend on being able to culture such viruses. However, the experience with coronaviruses has amply demonstrated that any systematic investigation of the viral ecology of domestic or wild animals is certain to yield many novel viruses. The question then is how to prioritize such information in terms of public health.

Ecological factors that predispose to increased proximity between donor and recipient populations, increases in their numbers, or promotion of interactions between different species within close confines (e.g. the live animal market trade) are likely to encourage interspecies transmission of viruses (Webster 2004). Thus, it may be intuitive in the first instance to focus on those animal species with greater opportunity for interaction with humans including those species that are part of the pet trade, domestic livestock, animals commonly sold for human consumption in 'wet-market' or live animal market settings and those commonly sold as 'bush-meat'.

In general, the more genetically variable and hence adaptable a virus is, the more likely is it to jump across species. Since RNA viruses are in general more genetically variable than DNA viruses, it is no surprise that the majority of emerging viral infections are caused by RNA viruses (Holmes & Rambaut 2004) making them the target of particular interest (Woolhouse et al 2005). It is also important to understand better the factors that predispose to or restrict inter-species transmission of viruses. Systematic studies of previous examples of disease emergence (e.g. SARS, past influenza pandemics) are likely to provide clues to the ecological and evolutionary changes that contributed to such emergence, so that we may be able to derive predictive algorithms for identifying those viruses in wildlife and domestic livestock that pose the greatest threat. For example, with increased realization about the increased diversity of bat coronaviruses, could we have predicted that the SARS-like bat coronaviruses constituted a public health risk? As in the fictitious thriller-movie, the 'Minority Report', is it feasible to identify the 'criminal' before the 'crime' is committed? This indeed is what is being attempted with influenza pandemic preparedness.

The highly pathogenic avian influenza virus H5N1 is currently regarded as the pre-eminent pandemic threat. This is based on the facts that (a) this virus has become entrenched in poultry populations in many countries, enhancing the opportunity for human exposure, and (b) that zoonotic transmission to humans continues to occur, albeit inefficiently (see above). In addition, there are many examples of the H5N1 virus crossing the inter-species barrier to infect other non-human mammalian species. It should be noted, however, that the influenza pandemics of 1957 and 1968 did not have the multi-basic cleavage site motifs in the connecting peptide of the virus haemagglutinin that is known to be associated with highly pathogenic avian influenza viruses. Pathogenicity in chickens is not therefore a reliable predictor of human pandemic risk.

What then may we use to assess the pandemic risk posed by an avian influenza virus? These include the opportunity for repeated interspecies transmission to humans, evidence of repeated zoonotic transmission to humans, adaptation of the virus to the relevant 'human' cell receptors, viruses undergoing rapid genetic change (reassortment), and adaptation of the virus to pigs (the hypothetical mixing vessel for reassortment with a human influenza virus). All these may be plausible criteria for pandemic concern (summarized in Table 1). While H5N1 clearly fulfils many of these criteria, it is not the only avian influenza virus to do so. H9N2 influenza viruses are even more widespread in poultry on a global scale than H5N1. They are undergoing rapid evolutionary change within domestic poultry (Xu et al 2007) and have repeatedly been isolated from humans with 'flu-like' illness (Peiris et al 2007). H6N1 subtype influenza viruses are another virus subtype that is commonly isolated from quail and other small poultry within southern China, and is also undergoing rapid evolution although it has so far not been isolated from humans (Cheung et al 2007).

However, as H9N2 virus is a low pathogenic avian influenza virus and does not commonly cause overt illness in chickens (other than a reduction of egg production on occasion) its presence usually goes unnoticed and unreported. Zoonotic transmission of H9N2 influenza in humans, when it occurs, is associated with a mild flu like illness rather than the severe and fatal disease associated with H5N1 infection (Peiris et al 2007). Therefore such human infections are rarely detected and, indeed, it is likely that there are more instances of zoonotic transmission of H9N2 to humans that of its more notorious cousin, H5N1. Further, the H9N2 virus currently endemic in poultry has a receptor-binding profile that allows binding to

TABLE 1 Hypothetical risk factors indicative of pandemic emergence

Hypothetical risk factors indicative of pandemic emergence	*H5N1*	*H9N2*	*H6N1*
Endemic in poultry over wide geographic area	++	+++	+
Genetic reassortment and rapid viral change	++	++	+
Evidence of repeated transmission to humans	+	+	No
Evidence of infection of pigs	+	+	No
Binding to receptors found in the human respiratory tract	?	+	No

both so-called 'human' α2'3 and 'avian' α2'6 receptors (Matrosovich et al 2001). Taken overall, one can argue that the H9N2 virus is at least as likely to cause the next pandemic as H5N1. Thus, the reason for the current focus on H5N1 as the next pandemic candidate is not its inevitability as such but rather the likely severity of such a pandemic, i.e. this is a low probability–high impact event.

In conclusion, it is clear that even with influenza, our understanding of the biological basis of inter-species transmission and the barriers that prevent or permit such transmission remains poorly understood. We have even further to go with many of the other viruses that are certain to be increasingly discovered in domestic livestock or wildlife if we are to make reliable predictions of zoonotic risk. However, recognition of our limitations is not a good enough reason not to attempt to achieve a better understanding of the biological determinants that govern interspecies transmission and zoonotic risk.

Acknowledgements

We acknowledge support from the National Institutes of Health, USA (NIAID contract HHSN266200700005), a Central Allocation Grant from the Research Grants Council of Hong Kong HKU1/05C and a grant from The Wellcome Trust, UK.

References

Beigel JH, Farrar J, Han AM et al 2005 Avian influenza A (H5N1) infection in humans. N Engl J Med 353:1374–1385

Bell D, Roberton S, Hunter PR 2004 Animal origins of SARS coronavirus: possible links with the international trade in small carnivores. Philos Trans R Soc Lond B Biol Sci 359:1107–1114

Chen H, Smith GJ, Zhang SY et al 2005 Avian flu: H5N1 virus outbreak in migratory waterfowl. Nature 436:191–192

Cheung CL, Vijaykrishna D, Smith GJ et al 2007 Establishment of influenza A virus (H6N1) in minor poultry species in southern China. J Virol 81:10402–10412

Ducatez MF, Olinger CM, Owoade AA et al 2006 Avian flu: multiple introductions of H5N1 in Nigeria. Nature 442:37

Fraser C, Riley S, Anderson RM, Ferguson NM 2004 Factors that make an infectious disease outbreak controllable. Proc Natl Acad Sci USA 101:6146–6151

Gu J, Xie Z, Gao Z et al 2007 H5N1 infection of the respiratory tract and beyond: a molecular pathology study. Lancet 370:1137–1145

Guan Y, Shortridge KF, Krauss S, Webster RG 1999 Molecular characterization of H9N2 influenza viruses: were they the donors of the "internal" genes of H5N1 viruses in Hong Kong? Proc Natl Acad Sci USA 96:9363–9367

Guan Y, Zheng BJ, He YQ et al 2003 Isolation and characterization of viruses related to the SARS coronavirus from animals in Southern China. Science 302:276–278

Hatta M, Hatta Y, Kim JH et al 2007 Growth of H5N1 influenza A viruses in the upper respiratory tracts of mice. PLoS Pathog 3:e133

Holmes EC, Rambaut A 2004 Viral evolution and the emergence of SARS coronavirus. Philos Trans R Soc Lond B Biol Sci 359:1059–1065

Hulse-Post DJ, Sturm-Ramirez KM, Humberd J et al 2005 Role of domestic ducks in the propagation and biological evolution of highly pathogenic H5N1 influenza viruses in Asia. Proc Natl Acad Sci USA 102:10682–10687

Ito T, Kawaoka Y 2000 Host-range barrier of influenza A viruses. Vet Microbiol 74:71–75

Kuiken T, Leighton FA, Fouchier RA et al 2005 Public health. Pathogen surveillance in animals. Science 309:1680–1681

Kung NY, Guan Y, Perkins NR et al 2003 The impact of a monthly rest day on avian influenza virus isolation rates in retail live poultry markets in Hong Kong. Avian Dis 47:1037–1041

Kung NY, Morris RS, Perkins NR et al 2007 Risk for infection with highly pathogenic influenza A H5N1 virus in chickens, Hong Kong, 2002. Emerg Infect Dis 13:412–418

Lau SK, Woo PC, Li KS et al 2005 Severe acute respiratory syndrome coronavirus-like virus in Chinese horseshoe bats. Proc Natl Acad Sci USA 102:14040–14045

Li KS, Guan Y, Wang J et al 2004 Genesis of a highly pathogenic and potentially pandemic H5N1 influenza virus in eastern Asia. Nature 430:209–213

Li W, Wong SK, Li F et al 2006 Animal origins of the severe acute respiratory syndrome coronavirus: insight from ACE2-S-protein interactions. J Virol 80:4211–4219

Lipsitch M, Cohen T, Cooper B et al 2003 Transmission dynamics and control of severe acute respiratory syndrome. Science 300:1966–1970

Matrosovich MN, Krauss S, Webster RG 2001 H9N2 influenza A viruses from poultry in Asia have human virus-like receptor specificity. Virology 281:156–162

Mounts AW, Kwong H, Izurieta HS et al 1999 Case-control study of risk factors for avian influenza A (H5N1) disease, Hong Kong, 1997. J Infect Dis 180:505–508

Nicholls JM, Chan MCW, Chan WY et al 2007 Tropism of avian influenza A (H5N1) in the upper and lower respiratory tract. Nat Med 13:147–149

Peiris JSM, Chu CM, Cheng VCC et al 2003 Clinical progression and viral load in a community outbreak of coronavirus-associated SARS pneumonia: a prospective study. Lancet 361: 1767–1772

Peiris JSM, Guan Y, Yuen KY 2004 Severe acute respiratory syndrome. Nat Med 10:S88–S97

Peiris JSM, de Jong MD, Guan Y 2007 Avian influenza virus (H5N1): a threat to human health. Clin Microbiol Rev 20:243–267

Poon LL, Chu DK, Chan KH et al 2005 Identification of a novel coronavirus in bats. J Virol 79:2001–2009

Shinya K, Ebina M, Yamada S, Ono M, Kasai N, Kawaoka Y 2006 Avian flu: influenza virus receptors in the human airway. Nature 440:435–436

Shortridge KF, Stuartharris CH 1982 An influenza epicenter. Lancet 2:812–813

Sims LD, Guan Y, Ellis TM et al 2003 An update on avian influenza in Hong Kong 2002. Avian Dis 47:1083–1086

Smith GJD, Naipospos TSP, Nguyen TD et al 2006 Evolution and adaptation of H5N1 influenza virus in avian and human hosts in Indonesia and Vietnam. Virology 350:258–268

van der Hoek L, Pyrc K, Jebbink MF et al 2004 Identification of a new human coronavirus. Nat Med 10:368–373

van Riel D, Munster VJ, de Wit E et al 2006 H5N1 virus attachment to lower respiratory tract. Science 312:399

Vijaykrishna D, Smith GJD, Zhang JX, Peiris JSM, Chen H, Guan Y 2007 Evolutionary insights into the ecology of coronaviruses. J Virol 81:4012–4020

Vong S, Coghlan B, Mardy S et al 2006 Low frequency of poultry-to-human H5NI virus transmission, southern Cambodia, 2005. Emerg Infect Dis 12:1542–1547

Webster RG 2004 Wet markets—a continuing source of severe acute respiratory syndrome and influenza? Lancet 363:234–236

Webster RG, Bean WJ, Gorman OT, Chambers TM, Kawaoka Y 1992 Evolution and ecology of influenza A viruses. Microbiol Rev 56:152–179

Woo PCY, Lau SKP, Chu CM et al 2005 Characterization and complete genome sequence of a novel coronavirus, coronavirus HKU1, from patients with pneumonia. J Virol 79: 884–895

Woolhouse ME, Haydon DT, Antia R 2005 Emerging pathogens: the epidemiology and evolution of species jumps. Trends Ecol Evol 20:238–244

Xu X, Subbarao K, Cox NJ, Guo Y 1999 Genetic characterization of the pathogenic influenza A/Goose/Guangdong/1/96 (H5N1) virus: similarity of its hemagglutinin gene to those of H5N1 viruses from the 1997 outbreaks in Hong Kong. Virology 261:15–19

Xu RH, He JF, Evans MR et al 2004 Epidemiologic clues to SARS origin in China. Emerg Infect Dis 10:1030–1037

Xu KM, Li KS, Smith GJD et al 2007 Evolution and molecular epidemiology of H9N2 influenza A viruses from Quail in southern China, 2000 to 2005. J Virol 81:2635–2645

Yamada S, Suzuki Y, Suzuki T et al 2006 Haemagglutinin mutations responsible for the binding of H5N1 influenza A viruses to human-type receptors. Nature 444:378–382

DISCUSSION

Osterhaus: In principle, I would share your conclusion about the fear for an H5N1 pandemic, but to a lesser extent than if it were a reassortment virus. If you get a virus that really mutates, for example, by adapting its receptor usage, I would be very scared of this. If it was a reassortment virus there would probably be a lot of pre-existing T cell immunity in the population at large. This might also be the difference between the 1918 virus on the one hand, and the 1957 and 1968 with two reassortments. The 1918 virus was greatly mutated so there is less immunity in the population at large.

Peiris: I would agree. Direct adaptation of the virus to human transmission as happened with 1918 is likely to be much worse than a reassorted virus. Having said this, we can't necessarily assume that reassortment would reduce the virulence, especially in regard to dissemination to multiple organs. With H5N1 disease in humans, there is some evidence that there is more dissemination than was thought up to now. I still would prefer to have an H9N2 pandemic than H5N1.

Osterhaus: I'm not sure you can say that. If H9 mutates into a virus that can easily transmit from human to human, a 1% case mortality rate would not be impossible. This would be a 1918 scenario.

Peiris: It's still better than H5N1. Even if H5N1 reassorts, I suspect it will still be a nasty pandemic.

Skehel: How widely spread are the H9N2 viruses in the avian populations?

Peiris: In domestic poultry H9N2 is widespread—even more so than H5N1. But H5N1 is catching up fast.

Skehel: How about the wild population?

Webster: It continues to be present in the wild bird population, but it is not the same virus as occurs in the domestic poultry population.

Skehel: This suggests that H9N2 is spread through commercial chains rather than through wild birds.

Peiris: It is highly endemic in poultry. It is generally asymptomatic or mild, and most of the time no one takes any notice of it.

Osterhaus: It causes disease, especially in animals that are pre-infected with IBV.

Skehel: To what extent are wild birds involved in the spread of H5N1?

Peiris: In the main centres of infection where it is endemic, wild birds are not playing a big role.

Osterhaus: In Vietnam, they swamped the area with avian vaccine, there were no more human cases, and then it relaxed: the same virus came back. This leaves the possibility that the virus was maintained in wild birds and not the domestic flock.

Webster: There was a group of ducks that did not respond to the vaccine administered: the muscovy ducks. The virus was in the muscovies all the time.

Peiris: In these endemic countries I think the wild birds are playing a relatively small role. On the other hand, in the movement of the virus from Qinghai Lake westwards, it has to be by wild birds. If it was through movement of poultry and poultry products, you'd have to explain why it happens to be this same virus that caused outbreaks in wild birds in Qinghai Lake, that was able to go on the Trans-Siberian railway and not any of the other H5N1 variants.

Osterhaus: I think it is probably both.

Webster: It was probably transmitted by wild birds but it is not being perpetuated in wild birds.

Osterhaus: It depends on how you define wild birds. In Niger we found it in vultures. They get infected because they eat a domestic chicken and then they transmit it to other vultures.

Webster: The internal genes in the early H5N1 probably came from an H9N2. The point I keep making is that in the early days in Hong Kong, the markets that were examined all had the virus. But there were no dead birds in those markets. There was probably T cell immunity based on H9N2. This H9N2 spread across the whole of Eurasia and is masking H5N1, and is responsible for the apparently healthy birds that are pooping out H5N1 that is highly pathogeneic.

Osterhaus: If what you are suggesting is true, would H9N2 be a good vaccine for humans?

Webster: There is a complex reactivity between H9N2 and H5N1 that is not worked out at the T cell level.

Peiris: Nina Kung who worked with us did some simulation market experiments. We created replica markets and infected birds with H9N2, looking at transmission. We had three different 'artificial' markets. One had H9N2-seronegative negative birds; one had a mixture of H9N2 seronegative birds and the other one had all

H5N1 vaccinated birds. Essentially there was no difference in transmission between the H5 vaccinated birds and the unvaccinated birds. The H5 vaccine doesn't seem to be interfering with transmission of H9N2. But of course, natural infection with H9N2 and consequent immunity will modulate H5 transmission while the vaccine may be different.

Wood: What is the extent of antigenic and genetic variation within H9N2? Is there a difference between the viruses in domestic and wild birds?

Peiris: Within domestic poultry there is a huge diversity. That virus is doing exactly the same as H5N1: it is rapidly reassorting, There are a number of lineages, and the viruses are also moving from ducks to chicken and back to ducks. They are doing exactly the same thing as H5N1 viruses.

Webster: There is a whole plethora of H9N2 viruses out there.

Smith: What about antigenic variation of the H9N2?

Peiris: There is some.

Wood: Are they still like A/Hong Kong/1073/99?

Webster: There are two dominant antigenic variants of H9N2.

Peiris: All of us are focusing on H5N1, but it is probably also important to pay more attention to H9N2.

Webster: It is interesting to go back to look at the history of H9N2. The virus that we know in Eurasia started not long before H5N1. It was the mid 1990s before the virus spread from the wild aquatic birds into domestic birds. This started in Asia and it became dominant in Korea. After that it spread to the whole of Eurasia. This preceded H5N1.

Osterhaus: I think this H9 story is important, but if we look at what comes out of the surveillance of wild birds, up to 20% may be infected. There are waves of infection, for example from H5 to H7. Even in the virtual absence of H9, if birds are being pre-infected with H5, there is a large exposure that may result in protection. Light infection with a low path H5 may be expected to give protection in the same way as you discussed for H9, against a highly pathogenic H5. I wouldn't focus only on H9.

Skehel: Does H9 go into pigs?

Peiris: Yes, we reported this back in 1998 (Peiris et al 2001). The Japanese have reported significant seropositivity of H9 in pigs (Ninomiya et al 2002).

Skehel: Is the residue at 226 the same in all the H9s? It is a leucine in the one we studied.

Peiris: Both lineages of H9 viruses have a capacity to bind human α2,6-sialic acid receptors.

Holmes: Are the clades of H5N1 always temporally and spatially distinct? Do you find more than one clade of H5N1 in the same place?

Peiris: In Indonesia there is just one clade. In Vietnam there are two, so they co-exist. But this is in a situation where infection has been suppressed quite a lot by very aggressive vaccination. In Thailand they used depopulation to suppress the poultry outbreaks. When they had re-emergent outbreaks last year in poultry there was the original clade 1 virus, but also some outbreaks caused by a different sublineage. These may be in a fairly artificial situation where the infection had been suppressed. In southern China, Guan Yi has reported geographically distinct sublineages in different parts of China. They seemed to be separated, except for regions such as Guangdong where there are multiple sublineages. This is not surprising since such regions have a lot of importation and exportation of poultry and are likely to be a melting pot of viruses. In a later study in China (Smith et al 2006) it appeared that this so-called geographical sublineages got wiped clean, and a different subclade (clade 2.3) has become dominant in the whole of the south China region. This virus has been found in Malaysia and Laos and also in some of the wild birds with H5N1 in Hong Kong.

Holmes: It sounds like the virus has been separated geographically into different lineages that have different fitness. It is not as if there is a progressive evolution in one place.

Smith: If you look phylogenetically, all of the clade twos look like they have a common ancestor. Having said that, if we look antigenically, it is a slightly different story. Antigenically it looks as if there is drift forced by either selection pressure or hitchhiking.

Osterhaus: Geographically, there is little competition. We don't see viruses taking over.

Holmes: Competition only takes place when you put them together.

Osterhaus: Aren't these areas being challenged back and forth?

Peiris: It may be that when you have a dominant virus already entrenched, it is more difficult for an outsider to come in.

Holmes: It is unlike humans where you have this big mixed population. There is more direction to it. In birds there is geographic separation and spatial distinctiveness, so the virus populations diverge by allopatric speciation. Then fitness plays a role when the populations come together.

Osterhaus: Humans live much longer and have less broad antibody responses than the birds. Humans have a more narrow response.

Skehel: Is that true? We use chicken sera to distinguish among viruses.

Smith: Adjuvanted sera doesn't distinguish anywhere near as well.

Hoffmann: How frequent is the transmission from domestic to wild birds? If the virus is established in migrating birds, it will be impossible to control H5N1 worldwide.

Peiris: In Hong Kong we do find wild birds with H5N1, and a lot of this may be being picked up from infected poultry. The spread to Europe must point to the fact that it is being maintained by wild birds.

Osterhaus: Originally, the idea was that we don't see these highly pathogenic viruses in wild birds. If there are sufficient numbers of wild birds together these viruses will arise anyway, but they are dead ends because they can not sustain themselves in wild birds. The populations aren't dense enough.

Smith: Highly pathogenic virus in one host isn't necessarily going to be highly pathogenic in another. There are now a sufficiently large number of species that can be infected with the H5 that you could have a low path strain that sustains itself.

Osterhaus: This is why we think Mallards are suspects for being real traffickers of the virus.

Peiris: There you have a problem with low path immunity, which may be allowing the birds to survive the infection.

Osterhaus: We just don't know how good that immunity is. If you have a mallard pre-infected by a low path H5N1 they are more or less vaccinated. Can they still excrete the high path virus?

Anderson: Is there evidence of antigenic drift or immune pressure?

Peiris: There is antigenic change, but this could be caused by vaccination.

Webster: There is a paper from David Suarez on the use of H5N2 vaccine in Mexico (Lee et al 2004). They followed H5N2 over many years and showed drift associated with the use of vaccine.

Osterhaus: I don't agree with the interpretation in that paper.

Smith: You mentioned the rest day in the markets in Hong Kong, and that this could break transmission in those markets. Is that just a rest day, or are the markets cleaned that day?

Peiris: They are cleaned as well. But normal cleaning alone without a rest day can't break transmission. Both are needed.

Osterhaus: This is an old veterinary principle. If you have pigs, farmers know never to have overlapping herds.

Webster: That's what the live poultry markets have been doing for years in Asia.

Lai: Bats seem to be the reservoir for so many different viruses. Why is this?

Anderson: They have high-density populations.

Holmes: That is the key issue. A dense population is good for harbouring pathogens. It's a strong ecological rule.

Osterhaus: As long as the virulence of the virus is low.

Holmes: The point is, the bigger the population is, the higher the virulence of the pathogen you can carry, because hosts are replenished quicker.

Osterhaus: There is also a selective advantage of having a population that can cover large distances and pick up the virus repeatedly.

Anderson: In addition to pathogenicity, ease of transmission comes into play.

Holmes: All these factors are related. A lot of work has been done on hunter–gatherer populations in humans, and they don't carry acute RNA viral infections endemically. They can get them, but the infection burns out. With measles, you don't see it endemically in populations of less than 300 000.

Osterhaus: This is why it isn't seen in monkeys because the populations are too small. But if monkeys come into contact with humans there are outbreaks of measles.

Smith: I'm not sure I can buy the idea that large populations can support highly pathogenic viruses. If you have a larger population you can get a huge epidemic spike, and it is no longer a large population. For a virus to be sustained in large populations it has to be low path in that group.

Holmes: The rule is that it is the rate at which you replenish your susceptible host, compared with the rate at which they are killed by virus. This has all been worked out mathematically.

Smith: If it were as pathogenic as H5 in chickens, it wouldn't matter how big that chicken flock was. But if a high path virus is to be reservoired in another species, you'd expect it to be lower path in the other species.

Osterhaus: In the bats, most of the viruses that they are reservoirs for are not pathogenic in them, such as rabies.

References

Lee CW, Senne DA, Suarez DL 2004 Effect of vaccine use in the evolution of Mexican linage H5N2 avian influenza virus. J Virol 78:8372–8381

Peiris JS, Guan Y, Markwell D, Ghose P, Webster RG, Shortridge KF 2001 Cocirculation of avian H9N2 and contemporary 'human' H3N2 influenza A viruses in pigs in southeastern China: potential for genetic reassortment? J Virol 75:9679–9686

Ninomiya A, Takada A, Okazaki K, Shortridge KF, Kida H 2002 Seroepidemiological evidence of avian H4, H5, and H9 influenza A virus transmission to pigs in southeastern China. Vet Microbiol 88:107–114

Smith GJD, Fan XH, Wang J et al 2006 Emergence and predominance of an H5N1 influenza variant in China. Proc Natl Acad Sci 103:16936–16941

Transmission and pathogenicity of H5N1 influenza viruses

Erich Hoffmann, Hui-Ling Yen, Rachelle Salomon, Neziha Yilmaz* and Robert G. Webster

*Department of Infectious Diseases, St. Jude Children's Research Hospital, Memphis, Tennessee 38105, USA and * Department of Virology, Refik Saydam Hygiene Institute, Cemal Gursel cad. No 18 Sihhiye, Ankara, 06100 Turkey*

Abstract. The interaction between multiple viral and host factors determine the pathogenicity and transmissibility of H5N1 influenza viruses. The viral surface glycoprotein haemagglutinin (HA) plays a crucial role in attachment to the host-cell sialic acid (SA) receptor and in viral growth and tissue tropism. Other viral factors known as host range or virulence determinants include viral polymerase, particularly residue 627 in the PB2 protein, and the ability to evade the host immune response through the viral NS1 protein. Applying plasmid-based reverse genetics, we have provided a good model system to study the molecular basis for pathogenicity. Differences in the pathogenicity in mammalian species were previously observed between human (A/Vietnam/1203/04) and chicken (A/Chicken/Vietnam/C58/04) H5N1 isolates. Detailed molecular characterization by plasmid-based reverse genetics showed that the polymerase subunits PB1 and PB2 contribute to the lethality of A/Vietnam/1203/04. The polymerase complex possessed significantly higher transcription/replication activity than the A/Chicken/Vietnam/C58/04 polymerase complex. Our results suggest that high polymerase activity acts as a molecular determinant for virulence in mammalian species. We also evaluated the transmissibility and pathogenicity of H5N1 viruses isolated from humans during the years 2003–2006 using a ferret contact model comprising one inoculated and two contact ferrets. This animal model has been shown to support efficient transmission of seasonal human influenza viruses. At 10^3 $TCID_{50}$, A/Vietnam/1203/04 and A/Vietnam/JP36-2/05 viruses, which have 'avian-like' α2,3-linked SA receptor affinity, caused neurological symptoms and death in ferrets. A/HongKong/213/03 and A/Turkey/65-596/06 viruses, which have affinity for both 'human-like' (α2,6-linked) and 'avian-like' SA receptors, caused mild clinical symptoms and were not lethal to ferrets. No transmission of A/Vietnam/1203/04 and A/Turkey/65-596/06 viruses was detected, and transmission of A/HongKong/213/03 and A/Vietnam/JP36-2/05 viruses was inefficient. These results demonstrate that despite their receptor binding affinity, circulating H5N1 viruses retain molecular determinants that restrict their spread among mammals.

2008 Novel and re-emerging respiratory viral diseases. Wiley, Chichester (Novartis Foundation Symposium 290) p 128–140

In domestic birds, such as chicken or turkeys, highly pathogenic avian influenza viruses (HPAIV) of the H5 subtype spread systemically, causing death within a few days. In 2004, re-emergence of H5N1 avian influenza in mammalian species was reported during severe poultry outbreaks in Asia. Fatal cases of human infection were identified in Vietnam, Thailand, Cambodia, and Indonesia. The spread of H5N1 viruses to Europe and Africa increased concern that an avian virus of Asian origin might gradually cross the species barrier and initiate an influenza pandemic. Avian influenza viruses may acquire the changes necessary to cause a devastating pandemic without genetic reassortment with human influenza viruses, as demonstrated by the likely avian origin of all eight gene segments of the 1918 pandemic influenza virus. Isolation of highly pathogenic H5N1 virus from sick or dead migrating birds is cause for concern, because their broad geographic range increases the likelihood that other species, including mammals, will be infected. There is limited knowledge about the factors that are important for virulence, adaptation, and transmission, and about how they are interrelated.

Human H5N1 influenza is characterized by viral pneumonia with acute respiratory distress syndrome (ARDS), diarrhoea, liver dysfunction and one reported case of central nervous system involvement. The severity of the disease and aspects of its pathogenesis differ from those associated with the human influenza viruses H3N2 and H1N1. While aspects of H5N1 virulence may reflect greater dissemination to multiple organs, including the central nervous system, most patients succumb to progressive primary viral pneumonia, ARDS, or other pulmonary compromise. Marked lymphopaenia is another consistent observation in patients with severe H5N1 disease and is an indicator of poor prognosis (Yuen et al 1998, Peiris et al 2004, Beigel et al 2005).

Viral factors responsible for H5N1 virulence in humans are not well characterized. A multiple basic amino acid motif at the HA cleavage site is important for virus dissemination and systemic spread in chickens. However, it is not known whether other adaptive changes in HA, such as those that determine receptor specificity, are crucial for tissue tropism and interspecies transmission. Typical avian strains have HAs that bind selectively to $\alpha(2,3)$-linked SA receptors, whereas human strains bind with greater affinity to $\alpha(2,6)$-linked SA. Less well understood is the role of internal gene segments. Polymerase subunit PB2 protein and non-structural NS1 protein were reported to be important factors in the high virulence of 1997 H5N1 viruses in mouse and pig models, respectively (Hatta et al 2001, Seo et al 2002, Shinya et al 2004).

Studies in animal models provide valuable information regarding the mechanism of pathogenesis of HPAIV in humans. Ferrets are considered an excellent small-animal model of human influenza. They exhibit clinical signs of infection similar to those observed in humans, such as fever, sneezing, diarrhoea, lethargy and acute respiratory illness (Zitzow et al 2002). Mice are also often used in influenza studies. Although human influenza viruses typically do not replicate in mice without adaptation, highly pathogenic H5N1 viruses isolated since 1997 can infect and kill mice without prior adaptation. Studies in these small-animal models provide valuable information about the contribution of viral genes to pathogenesis and transmission. Recent technological advances allow us to reconstitute entire influenza virus genomes, thereby significantly improving our ability to determine the molecular basis of virus adaptation, transmission, and virulence. Here we summarize our recent findings on viral genes associated with the high pathogenicity of an H5N1 strain isolated from a fatal human case and on the transmissibility of H5N1 viruses isolated during the years 2003–2006.

Results

Viral genes associated with high pathogenicity in mammals

Our group and others have demonstrated that the H5N1 viruses A/Vietnam/1203/04 (VN1203), isolated from a fatal human case, and A/Chicken/Vietnam/C58/04 (CH58) are both highly pathogenic to chickens. However, only the human isolate is lethal to ferrets (Govorkova et al 2005, Maines et al 2005). To determine what gene segments contribute to the different virulence and systemic spread of these viruses, we used the eight-plasmid reverse genetics system (Hoffmann et al 2000a.) We constructed two eight-plasmid sets encoding the individual genes of VN1203 and CH58 and generated reverse genetics (RG) VN1203 and CH58 recombinant viruses by DNA transfection (Fig. 1). Protein sequences of the RG viruses were identical to their respective parental viruses. To test the pathogenicity of the RG viruses, we intranasally inoculated ferrets with 10^6 50% egg infectious doses (EID_{50}) of virus. VN1203 virus was lethal to ferrets, but CH58 was not. The ferrets' body temperatures were approximately 2 °C higher after inoculation with VN1203 than after inoculation with CH58. Ferrets inoculated with VN1203 lost more than 20% of their body weight, while those in the CH58 group maintained or increased their weight. These findings were similar to those observed after inoculation with the respective parental virus strains (Govorkova et al 2005).

The systemic spread and high lethality of VN1203 in ferrets and mice could be caused by altered tissue tropism resulting from changes in the HA and neuraminidase (NA) surface glycoproteins. VN1203 and CH58 differ by six amino acids in each glycoprotein. We therefore generated a RG VN1203-CH58(HA,NA) virus composed of the CH58 HA and NA genes plus the six internal genes of

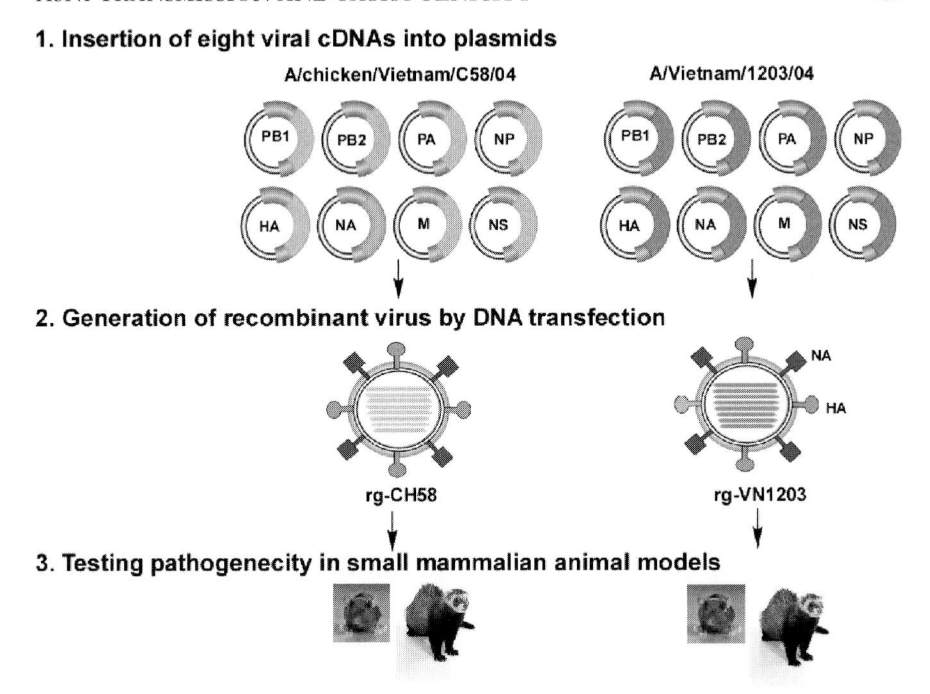

FIG. 1. Model system used to study the molecular basis of pathogenicity in different mammalian hosts. Each of the plasmids contains an individual gene segment from a low-pathogenic H5N1 virus (A/chicken/Vietnam/c58/04) and a high-pathogenic human H5N1 isolate (A/Vietnam/1203/04). The genes important for virus spread in mammalian host species are identified by mixing the appropriate plasmids to generate recombinant viruses for testing in small mammalian animal models.

VN1203. All ferrets inoculated with this recombinant virus died between days 6 and 11 post-inoculation (p.i.), and all inoculated mice died by day 8 p.i. The survival rate, weight loss, virus titres, and disease severity caused by VN1203-CH58(HA,NA) was similar to that caused by VN1203 in both mammalian models.

The polymerase genes of VN1203 and CH58 differ by 11 amino acids (4 in PB2, 3 in PB1, and 4 in PA). Inoculation with the reassortant virus VN1203-CH58(3P), containing the CH58 PB2, PB1 and PA genes, was not lethal to ferrets or mice and caused no weight loss. Ferrets inoculated with the single-gene reassortant VN1203-CH58(PB2) experienced modest weight loss, and 2 of the 3 inoculated ferrets survived. Nasal wash virus titres were 2 logs lower on day 1 and 1 log lower on day 5 in those inoculated with reassortant virus containing CH58 PB2 rather than VN1203 PB2. Mice inoculated with VN1203-CH58(PB2) showed no weight loss, and 1 of 8 mice died, 10 days p.i. A RG

VN1203 virus containing a point mutation that substituted glutamic acid (E) for lysine (K) at position 627 of PB2 was not lethal to mice. These results demonstrate that K627 in VN1203 PB2 contributes to lethality in mice. Half of the ferrets inoculated with single-gene reassortant VN1203-CH58(PB1) survived. Mice inoculated with the same reassortant showed a 15% weight loss 7 days p.i., but all mice gained weight after day 9 and survived. These results underscore the importance of genetic changes in PB2 and PB1 to the replication efficiency of VN1203.

To test whether the three VN1203 polymerase genes are sufficient for high lethality, we generated a reassortant virus comprising the human isolate's polymerase genes and the remaining five segments from the chicken isolate. Inoculation of mice with CH58-VN1203(3P) resulted in severe weight loss and 100% lethality between days 9 and 11 p.i. Two of three ferrets inoculated with this virus survived and lost more than 12% body weight. Taken together, these results demonstrate that the polymerase genes contribute to pathogenicity in both mice and ferrets.

To compare the polymerase activity of VN1203 and CH58, we applied influenza A virus minigenome assay by transfecting 293T cells with plasmids containing their PB2, PB1, PA and NP genes and a luciferase reporter plasmid. Because cDNA encoding luciferase is controlled by the influenza A virus M segment's non-coding region, luciferase levels reflected overall transcription and replication activity of the polymerase complex (Hoffmann et al 2000b). Luciferase activity (relative light units; RLUs) in cells co-transfected with VN1203 polymerase complex and NP was 3.5-fold higher than that in cells co-transfected with CH58 polymerase complex and NP. Therefore, the VN1203 polymerase complex and NP showed higher transcription/replication activity than the CH58 polymerase complex and NP. To assay the individual effect of VN1203 PB2 and PB1 on polymerase activity, we replaced the individual PB2 and PB1 segments of VN1203 with those of CH58. When VN1203-PB2 was replaced with CH58-PB2, luciferase activity was reduced by 78%. When VN1203-PB1 was replaced with CH58-PB1, luciferase activity was reduced by 42%. The effect of CH58 NP was also assayed by co-transfecting VN1203 PB2, PB1, and PA plasmids with the CH58 NP plasmid. The luciferase activity did not differ from that observed with the VN1203 PB2, PB1, PA and NP plasmids. Overall, the VN1203 polymerase complex and NP showed significantly higher transcription/replication activity than the CH58 polymerase complex and NP, and VN1203 PB2 and PB1 may have contributed to the greater polymerase activity observed from VN1203 polymerase complex and NP.

The NS1 protein is known to regulate innate immunity by modulating the host type I interferon response, which has both antiviral and immunoregulatory functions. We therefore generated single-gene reassortants that combined CH58 NS with the remaining seven VN1203 genes. Five of six ferrets inoculated with the reassortant virus survived. They lost 20% less weight than did ferrets inoculated with VN1203. All mice inoculated with this reassortant lost weight and died by day 8 p.i. Thus, the contribution of NS to lethality differs between ferrets and inbred mice.

Evaluation of the transmissibility and pandemic potential of H5N1 viruses

H5N1 virus that re-emerged in late 2003 can be grouped into several genetic clades based on phylogenetic analysis of the HA gene and viruses that belong to different clades/sub-clades are antigenically distinguishable by HI assay. To obtain a broad representation of the clades of circulating H5N1 viruses, we selected human isolates from clades 1 [A/HongKong/213/03 (HK213), A/Vietnam/1203/04 (VN1203), and A/Vietnam/JP36-2/05 (VNJP36-2) viruses] and 2 [A/Turkey/65-596/06 (TK65596) virus]. Three of these isolates contained amino acid changes in the conserved HA receptor-binding residues. The HK213 and TK65596 viruses contained N rather than the conserved S at residue 223; VNJP36-2 virus contained A rather than the more conserved S at 133.

We measured the receptor binding affinity of the four viruses to high-molecular-weight sialic substrates. All isolates exhibited affinity for p3′SL (synthetic sialosaccharides with α2,3-linkage). The HK213 and TK65596 viruses, both of which had an S→N substitution at HA receptor binding residue 223, showed a greater affinity for p6′SL (synthetic sialosaccharides with α2,6-linkage) than is typical of avian H5N1 influenza viruses, and both of these viruses had an S→N substitution at HA receptor binding residue 223. These results are consistent with previous reports (Hoffmann et al 2005, Shinya et al 2005) that N at residue 223 of HA may alter the receptor binding affinity of the H5 HA glycoprotein.

We showed previously that our ferret contact model is suitable for transmissibility studies of human H3N2 viruses (Yen et al 2005). Applying the same experimental design, we assessed the transmissibility of each H5N1 virus in two independent ferret groups (the TK65596 virus was studied in only one ferret group). No significant difference was observed between nasal wash titres of the four viruses 2, 4, or 6 days after inoculation with 10^3 $TCID_{50}$. Nasal wash titres usually peaked at 4 days p.i. (range, 2–6 days), and peak titres were 4.0–6.5 $log_{10}TCID_{50}$. Between days 2 and 6 p.i., virus (1.7–2.5 $log_{10}TCID_{50}$) was also detected in the rectal swabs of ferrets inoculated with VN1203 and VNJP36-2 viruses.

Transmission of virus from inoculated donors to naïve contact ferrets was confirmed by isolation of virus from nasal washes or rectal swabs and/or by serologic testing. No transmission of the VN1203 or TK65596 virus was detected. One of four contact ferrets exposed to HK213 had developed neutralizing antibody (titre = 1 : 80) when tested after 21 days of contact with the inoculated ferret but did not shed detectable virus or show signs of illness. After re-challenge with HK213 virus, this contact ferret did not shed virus, whereas the seronegative (neutralizing antibody titre $<1:20$) contact ferrets did. These results showed that HK213 virus is not transmitted efficiently between ferrets and appears not to cause symptomatic infection after transmission.

In each of two independent ferret groups, VNJP36-2 virus was isolated from one of the two naïve contact ferrets after 6 and 8 days, respectively, of contact with the inoculated ferret (8 and 10 days after inoculation of the inoculated ferret). One contact ferret in each group remained seronegative after 21 days of contact, demonstrating the inefficiency of VNJP36-2 transmission and the absence of secondary transmission. Despite high virus titres in the nasal washes of inoculated ferrets, little or no virus was detected in the nasal washes of the infected contact ferrets. VNJP36-2 virus was detected on only one day in the nasal wash of one infected contact; virus was detected in the rectal swabs of both infected contact ferrets on one day.

The four H5N1 viruses were less efficiently transmitted in this animal model than the human H3N2 A/Wuhan/359/95 virus, which was transmitted efficiently from the inoculated ferret to both naïve contact ferrets after 2 to 4 days of contact (Yen et al 2005). However, we observed differences in transmissibility between the four H5N1 viruses in this animal model. The VN1203 and TK65596 viruses were not transmitted to contact ferrets; the HK213 virus was transmitted to one contact ferret but was detectable only by seroconversion; and the VNJP36-2 virus was transmitted with a significant delay compared to human H3N2 virus to one of two contact ferrets in two separate ferret groups, as shown by virus shedding and/or seroconversion.

The four human H5N1 isolates differed in their pathogenicity in inoculated ferrets. VN1203 and VNJP36-2 were lethal and caused neurological signs, temperature elevation (maximum elevation, 2.4–3.7 °C), and weight loss (11%–29%) similar to the disease signs previously reported (Govorkova et al 2005, Maines et al 2005). In contrast, ferrets inoculated with the HK213 and TK65596 viruses showed only slight temperature elevation (0.9–1.4 °C) and weight loss (2.4%–9.7%). To evaluate virus replication and tissue tropism in the inoculated ferrets, we titrated virus 5 days p.i. in the nasal turbinate, trachea, lungs, brain, olfactory bulb and large intestine of two ferrets per virus. For comparison, we inoculated one ferret with 10^3 $TCID_{50}$ of recombinant Wuhan/95 (H3N2) virus. All four human H5N1 isolates were detected in the olfactory bulb, nasal turbinate, lungs, and large intestine; however, only the VN1203 and VNJP36-2 viruses were detected in the brain. In contrast, the H3N2 human influenza virus was detected only in the nasal turbinate. Of the four H5N1 viruses, VN1203 replicated most efficiently in all ferret organs except the trachea.

Discussion

Plasmid-based reverse genetics now allows us to map the factors that contribute to the virulence of H5N1 influenza virus isolates. We showed that recombinant human influenza virus isolate VN1203, unlike CH58, rapidly spread systemically to the brain and liver in ferrets, resulting in high lethality. The survival of all mice and ferrets inoculated with a reverse genetics VN1203 virus that contained the polymerase genes of the chicken isolate showed that these genes are important for virulence. In the reverse experiment, a recombinant virus comprising the chicken-isolate backbone with the VN1203 polymerase genes killed all inoculated mice and one out of three inoculated ferrets. Applying influenza A virus minigenome assay, we observed that VN1203 polymerase subunits PB2 and PB1 have contributed to the high polymerase activity and likely the pathogenicity of the VN1203 virus. Since polymerase subunits PB2 and PB1 are central to the influenza virus replication cycle and are required for viral RNA replication and transcription, inhibition of this function by antivirals directed against the polymerase proteins may dramatically reduce the severity of disease and prevent death. Such new antivirals could complement the use of neuraminidase inhibitors and the M2-ion channel blocker amantadine.

The role of HA in virus cell binding and entry implies that it is important for efficient virus spread after infection. The presence of a multibasic cleavage site in the HA is well correlated with high pathogenicity in influenza A viruses. This motif allows cleavage by intracellular furin-like proteases. In this study, both the chicken and human isolate had polybasic amino acids in the connecting HA peptide, which allowed rapid systemic viral spread; this feature results in lethality to chickens and is essential for the virulence of the 1997 human H5N1 isolates in mice. However, the contribution of HA receptor-binding properties to cell tropism and virus transmission are not known. In our model system, both VN1203 and CH58 viruses were selective for α-(2,3)-linked SA and spread efficiently after experimental infection. It is noteworthy that CH58, VN1203, and most of the recent H5N1 viruses have S223 in their HA and that this residue, located in the HA 220-loop, is crucial for α-(2,3)-linked SA receptor specificity. Substitution of N for S at residue 223 is associated with changes in receptor binding affinity and specificity, as field H5N1 isolates HK213 and TK65596 possess affinity towards both synthetic sialyl substrates with the avian-like α-(2,3)-linkage and the human-like α-(2,6)-linkage. Our results on VN1203 and CH58 viruses highlight the contribution of PB1 and PB2 to high lethality in mammalian hosts, and that multibasic amino acids in the cleavage site of the HA are necessary but not sufficient for high lethality in a mammalian host.

The inefficient transmission of both the low-pathogenic HK213 and the high-pathogenic VNJP36-2 viruses in ferrets suggests that there is no direct correlation between pathogenicity and transmissibility of the H5N1 viruses in this animal model. Interestingly, however, the less pathogenic HK213 virus caused only asymptomatic infection in one contact ferret, while the more pathogenic VNJP36-2 virus caused severe clinical signs in infected contact ferrets. The inefficient transmission of HK213 and the non-transmission of TK65596 virus demonstrates that the observed binding affinity for 'human-like' α-(2,6)-linked SA receptors is not sufficient to allow efficient transmission of H5N1 virus among ferrets. It is likely that the avian-derived H5N1 viruses require further adaptation in their surface glycoproteins and internal genes to allow efficient co-operation with the mammalian cell machinery.

Acknowledgements

The authors thank the Refik Saydam Hygiene Institute in Turkey, Ahmet F. Oner, and the WHO Global Influenza Surveillance Network for providing the H5N1 viruses; Nicolai V. Bovin for the gift of sialic polymer substrates; the influenza group at St. Jude Children's Research Hospital for comments and support; and Sharon Naron for editorial assistance. This study was supported by grants AI95357 and CA21765 from the National Institutes of Health and by the American Lebanese Syrian Associated Charities (ALSAC).

References

Beigel JH, Farrar J, Han AM et al 2005 Avian influenza A (H5N1) infection in humans. N Engl J Med 353:1374–1385

Govorkova EA, Rehg JE, Krauss S et al 2005 Lethality to ferrets of H5N1 influenza viruses isolated from humans and poultry in 2004. J Virol 79:2191–2198

Hatta M, Gao P, Halfmann P, Kawaoka Y 2001 Molecular basis for high virulence of Hong Kong H5N1 influenza A viruses. Science 293:1840–1842

Hoffmann E, Neumann G, Kawaoka Y, Hobom G, Webster RG 2000a A DNA transfection system for generation of influenza A virus from eight plasmids. Proc Natl Acad Sci USA 97:6108–6113

Hoffmann E, Neumann G, Hobom G, Webster RG, Kawaoka Y 2000b Ambisense approach for the generation of influenza A virus: vRNA and mRNA synthesis from one template. Virology 267:310–317

Hoffmann E, Lipatov AS, Webby RJ, Govorkova EA, Webster RG 2005 Role of specific hemagglutinin amino acids in the immunogenicity and protection of H5N1 influenza virus vaccines. Proc Natl Acad Sci USA 102:12915–12920

Maines TR, Lu XH, Erb SM et al 2005 Avian influenza (H5N1) viruses isolated from humans in Asia in 2004 exhibit increased virulence in mammals. J Virol 79:11788–11800

Peiris JS, Yu WC, Leung CW et al 2004 Re-emergence of fatal human influenza A subtype H5N1 disease. Lancet 363:617–619

Seo SH, Hoffmann E, Webster RG 2002 Lethal H5N1 influenza viruses escape host anti-viral cytokine responses. Nat Med 8:950–954

Shinya K, Hamm S, Hatta M, Ito H, Ito T, Kawaoka Y 2004 PB2 amino acid at position 627 affects replicative efficiency, but not cell tropism, of Hong Kong H5N1 influenza A viruses in mice. Virology 320:258–266

Shinya K, Hatta M, Yamada S et al 2005 Characterization of a human H5N1 influenza A virus isolated in 2003. J Virol 79:9926–9932

Yen HL, Herlocher LM, Hoffmann E et al 2005 Neuraminidase inhibitor-resistant influenza viruses may differ substantially in fitness and transmissibility. Antimicrob Agents Chemother 49:4075–4084

Yuen KY, Chan PK, Peiris M et al 1998 Clinical features and rapid viral diagnosis of human disease associated with avian influenza A H5N1 virus. Lancet 351:467–471

Zitzow LA, Rowe T, Morken T, Shieh WJ, Zaki S, Katz JM 2002 Pathogenesis of avian influenza A (H5N1) viruses in ferrets. J Virol 76:4420–4429

DISCUSSION

Webster: So we are starting to see some information on the number of genes involved in transmissibility. This issue is obviously not straightforward.

Skehel: You mentioned the 223 mutation. Did you look for neuraminidase inhibitor resistance in those viruses?

Hoffmann: Not in this study.

Skehel: The initial studies on the Turkey virus indicated that the viruses that had the change at 223 were more resistant to Tamiflu. The implication from this might be that the haemagluttinin has a lower affinity, quite apart from the differences in specificity that you showed. The lower affinity could also compromise transmissibility.

Hoffmann: That is a good point. Low affinity binding may contribute. However, it appears that other genes in addition to the haemaggluttinin are important for transmission.

Lai: Your studies show that the molecular determinants for pathogenicity differ among different animals. Which animal model is more representative of the human pathogenicity?

Hoffmann: We had some discussion on this earlier: what is the best model system? It depends on the question one is asking. We can only answer this by testing different strains with different animal species. For example, in our studies we found that polymerase genes are important for pathogenicity in mice and ferrets. In this case we believe that the general mechanism is a more efficient virus growth early after infection that leads to the high lethality. Therefore, both mouse and ferret are excellent models.

Osterhaus: The obvious way to go is to do site-directed mutation in the individual, such as taking 223 and mutating it back, and then looking at what it does in the respective species. We did the same for H7, for the lethal virus and the mild virus.

To make all the different combinations is quite a bit of work. It also depends on the animal species you test it in.

Hoffmann: It's good that we have reverse genetics.

Osterhaus: There is pathogenicity on one hand and transmissibility on the other hand. Standardized assays for transmissibility are relatively difficult to perform. How do we define transmissibility in different species? Another complication is the behaviour of the different animals.

Webster: How many of these changes are showing up in the domestic animals?

Hoffmann: The only signature sequence we know of where we can relate mammalian influenza virus with avian influenza virus is at position 627 of the polymerase subunit PB2, which is lysine in mammals and glutamic acid for avian isolates. This is not 100%. A few viruses isolated from avian species have lysine at PB2-627.

Holmes: There is a whole clade of the bird viruses that has that mammalian amino acid.

Webster: Have you also found this in the H7 viruses?

Osterhaus: The highly pathogenic virus for humans and mice has a number of mutations that arose in chickens. This is difficult to understand if there wasn't a mammalian transmission in between. We went back to the farm that the chickens came from and found a number of mutations.

Peiris: Regarding transmissibility, the 223 substitution also converts a 2,3 virus to a 2,3 plus 2,6 dual binding virus. Yoshi also showed this change in receptor binding in recent human viruses from 2004. But they do not transmit efficiently in humans. Have you looked at going all the way to losing the 2,3 site, in addition to acquiring the 2,6 binding?

Hoffmann: I think you are referring to the neighbouring positions to 223 known to be important for receptor specificity that could be changed using reverse genetics.

Peiris: Is this a safety issue?

Webster: Biosecurity people aren't happy with this.

Skehel: Can you tell us anything about the mechanism of the 627 mutation being involved in polymerase activity. Does it influence interaction of the polymerase with host components? If so, which host components might be involved?

Hoffmann: I'd like to know that. The sequence evidence suggests that lysine at position 627 is related to the mammalian species. This suggests an interaction with a host factor. Which host components are involved is not known.

Osterhaus: You have to be careful with the term 'mammalian', because humans are different from cats, which are different from mice and so on.

Hoffmann: I agree. Referring to 'mammalian' factors in general is a simplification of a possibly much more complex underlying biology.

Lai: In *in vitro* polymerase assays, which cell types can we use? Do we have to use human cells, or ferret cells, or mouse cells?

Hoffmann: We use 293 T cells, which are human cells.

Lai: The polymerase activity might be different in different cell types.

Hoffmann: Yes. We are focusing on the comparison between polymerase activity in human cells and avian cells to obtain evidence of contributions of host factors or polymerase subunits *per se* contribute to higher polymerase activity.

Anderson: The polymerase complex changes. What other phenotypic data do you have on differences in lethality in the mouse and ferret that you saw? Does virus titre correlate with lethality?

Hoffmann: The highly virulent H5N1 viruses spread systemically in the ferret and mouse resulting in high titres in multiple organs.

Thiery: Have you looked at any semi-quantitative methods for measuring the K_{on} and K_{off} of this interaction?

Hoffmann: Virus binding affinity to sialylglycopolymers obtained by conjugation of a 1-*N*-glycyl derivative of 3′-sialyllactose or 6′-sialyllactose was tested by competitive assay based on inhibition of binding to peroxidase-labelled fetuin.

Thiery: You need a biacore or plasmon resonance method to measure the K_{on} and K_{off} of the interaction.

Hoffmann: That would be interesting to do.

Lal: Do you know which component, PB1 or PB2, is responsible for slowing down the activity of the polymerase? There may be just one component. If you used a siRNA to slow down each one of these complexes you might be able to pin down which is involved?

Hoffmann: Both contribute. PB1 is the subunit that catalyses the elongation of the viral RNA chains, and PB2 is the protein that 'steals' the cap-structure from cellular mRNAs.

Osterhaus: You said quite categorically that α2,3/α2,6 is not the whole story about transmissibility. What are we missing?

Hoffmann: I think that we do not understand the contribution of the other gene segments to transmissibility. Possibly, we are missing some important mechanisms occurring in the upper respiratory tract that allow some viruses such as H3N2 to easily transmit, but other even more efficiently growing viruses such as H5N1 viruses not to transmit.

Skehel: It isn't just binding specificity of HA; it is affinity as well. There is evidence that the change at HA1 223 influences affinity from the neuraminidase sensitivity data. If change in the HA makes it less susceptible to neuraminidase inhibitors, this could be because mutation in the HA has led to the change in affinity because the virus then gets off the infected cell surface more easily without having to have the neuraminidase working efficiently.

Osterhaus: How does that translate into transmissibility? Is it availability of the virus?

Skehel: If the receptor binding affinity of HA is lower then the transmissibility it will probably be lower.

Hoffmann: 'Loose' binding may be important for more efficient release and spread.

Lai: We think of transmissibility as the infection of different host species. Transmissibility may also mean spread of the virus between different tissues. Have you looked at the spread of the virus in different organs?

Hoffmann: Our definition for transmission is between different animals. The virus clearly spreads to multiple organs, including the brain and intestine. Thus, the virus grows efficiently inside the host, but there appears to be a 'block' for spreading of H5N1 viruses between individual hosts.

Anderson: Do the PB1 and PB2 mutations affect transmission?

Hoffmann: We don't know. The studies we have done on the polymerase genes have solely concerned pathogenicity, with experimental infection of ferrets.

Peiris: Talking about the reduced affinity interaction of the virus, this must mean that the viruses are less efficient at infecting. Is there any evidence that the 213 virus that carries the 223 mutation gives lower titres in animals?

Hoffmann: It grows to lower titres, but we did not test the contribution of other genes.

Peiris: If you have a low affinity receptor interaction, it should translate into many things, not just transmissibility.

Kahn: Is the virus that comes from these animals infective?

Osterhaus: Yes. There is plenty of virus and it is infective.

Webster: It is the same in the pig: the virus is present in sufficient quantity.

Skehel: Do the pigs get sick with H5?

Webster: Not with the H5 that we have inoculated into pigs experimentally, but there are reports in the field of sick pigs. It is a complex issue. The missing link in whether this virus will be successful or not in transmissibility.

Development of vaccine for a future influenza pandemic

John M. Wood

National Institute for Biological Standards and Control, Blanche Lane, South Mimms, Potters Bar, Herts EN6 3QG, UK

Abstract. Vaccination offers the most effective large scale preventative measure against pandemic influenza, but experience gained from development of H5N1 vaccines over the past four years has shown that we are not yet ready to react sufficiently quickly and effectively to obtain the most benefit from pandemic vaccines. Although safe and productive vaccine viruses can be prepared from highly pathogenic H5N1 viruses by the use of reverse genetics technology, vaccine is unlikely to be ready in time for the first pandemic wave. Another issue facing us is the use of hens' eggs for vaccine production. During a pandemic, the supply of eggs may be compromised and it is important to move ahead with vaccine production based on use of mammalian cells. We have found that vaccine strategies designed for seasonal influenza are unlikely to stimulate protective immunity in pandemic situations, so alternative strategies need to be explored. Finally the regulatory authorities face challenges in licensing and testing new pandemic vaccines, and recent progress will be reviewed.

2008 Novel and re-emerging respiratory viral diseases. Wiley, Chichester (Novartis Foundation Symposium 290) p 141–151

It is generally accepted that vaccines will provide our best intervention strategy against pandemic influenza. In recent years, the episodes of avian influenza virus infection of humans suggest that the next pandemic virus may be a highly pathogenic avian strain, and we are faced with the problem of generating a vaccine from a potentially lethal virus. One way around this problem that became feasible only in 2003 is to use 'reverse genetics' to reliably create a safe vaccine virus. We know that the major molecular basis for virulence of H5N1 viruses is linked to the cleavability of the H5N1 haemagluttinin (HA) spike protein, in addition to important features of the internal viral proteins, and that the enhanced cleavability is due to an extra 4–6 basic amino acids in the HA structure at the cleavage site. Using genetic engineering techniques, these extra basic amino acids can be excised from a cloned copy of the HA segment. At the same time, by rescuing the neuraminidase (NA) segment from the potential pandemic strain along with the backbone of six gene segments from a human virus A/PR/8/34 (H1N1) (PR8), a virus can be generated that has the outer coat proteins of the pandemic strain, has the ability to grow well in eggs, and is non-pathogenic. The use of PR8 as a genetic backbone for pandemic reference viruses has been approved by the WHO.

Creation of a safe vaccine virus is only the start of pandemic vaccine development, however. There are several key questions that must be answered:

- Will vaccine be ready in time for a pandemic?
- Can we produce a pandemic vaccine without depending on eggs?
- Can we produce a vaccine that will be protective at low antigen content?
- Are the vaccine regulators ready to license the new pandemic vaccines?

Speed of vaccine virus development

There are three laboratories approved by WHO to produce pandemic vaccine viruses: St Jude Children's Research Hospital, USA, Centers for Disease Control, USA and NIBSC in the UK. The key facilities for such work include a high containment laboratory (BSL3 or BSL4 depending on locally implemented laws); experience with reverse genetics; access to qualified mammalian cells for virus rescue; and a quality system suitable for the early stages of vaccine development. Once the candidate pandemic vaccine virus has been produced, it is important to check that the virus is safe to send to vaccine manufacturers and a series of tests has been agreed by WHO for this purpose including *in vitro* and *in vivo* tests (WHO 2005). In the event of a pandemic, it is absolutely vital to prepare vaccine reference viruses as quickly as possible. In 2004, when we were faced with the threat of human H5N1 infections in Vietnam and Thailand, it took 2.6 months to produce a fully-tested vaccine virus (NIBRG-14) at NIBSC. Since that time, we have streamlined as much of the process as possible, with the result that our current capability is just 1.4 months. However, there is an opportunity to do even better, if we consider that after just one month it is possible to have a newly derived H5N1 reverse genetics virus, which has been tested for absence of multiple basic amino acids at the HA cleavage site, for the inability to kill chicken embryos and the inability to form plaques in mammalian cells without the addition of trypsin. We would thus have a great deal of confidence, based on our experience and that of others, that the candidate H5N1 vaccine virus is safe. The remaining tests for safety in ferrets and chickens, which need a further 0.5 months, would not have been completed. According to current OIE regulations, an H5N1 virus which has not been tested for pathogenicity in chickens is considered to be highly pathogenic (OIE 2005) and it would not be possible to distribute a partially-tested H5N1 virus unless it is handled at either BSL3 or 4. A saving of 0.5 months would be vital in preparing a pandemic vaccine and for this reason, the OIE has recently introduced emergency measures to allow such vaccine viruses to be released to vaccine manufacturers before the chicken pathogenicity test has been completed (WHO 2007). However in the USA, a partially-tested H5N1 virus is designated 'select agent' status by the USDA, which would impede the early release of a pandemic virus. Discussions are needed to relax such laws under emergency situations. Emergency legislation is now in place in the USA (note added in proof, December 2007).

If we assume that a pandemic vaccine virus can be made available to vaccine manufacturers within one month, will vaccine be ready in time? Unfortunately the vaccine manufacturers estimate that it will take a further 3–4 months before the first vaccine doses are ready, so if the next pandemic spreads around the world as quickly as anticipated, we are unlikely to have a vaccine until well into the first pandemic wave. There are some options open to us to improve pandemic vaccine availability including stockpiling vaccine; immunizing before a pandemic (immunological priming); and production of vaccine viruses in advance (a library of reagents). However with each of these strategies there is an element of risk because there is no way of knowing in advance, what the pandemic virus will be. However because of the consequences of not having vaccine in time, it may be worthwhile to take such a risk.

Dependence on eggs

It is generally acknowledged that rapid production of pandemic vaccines depends largely on availability of hens' eggs and that there is a great shortfall between the global influenza vaccine capacity (300 million doses of trivalent vaccine) and the global need for pandemic influenza vaccines (6.2 billion doses) (Fedson 2006). The situation is particularly acute in developing countries that do not have sufficient vaccine manufacturing capacity. In order to meet this shortfall, the World Health Assembly (2005) and various governments are encouraging the rapid development of vaccine based on mammalian cell culture. There are some distinct advantages in the use of cell culture, including the possibility to produce vaccine directly from highly pathogenic viruses under high containment, which will obviate the need for a reverse genetics virus thus saving time and independence from eggs, whose supply could be threatened by pandemic influenza caused by highly pathogenic avian viruses. Although our current pandemic vaccine capability depends on eggs, there is an opportunity to extend global influenza vaccine production principally due to increasing use of cell culture technology.

Pandemic vaccine efficacy

There is striking evidence from clinical trials of candidate pandemic vaccines prepared and delivered using conventional technology, that they are unlikely to be protective unless they are used at high antigen concentrations. In the EU, a subunit H5N3 vaccine prepared from a clade 3 A/Duck/Singapore/97 virus (Nicholson et al 2001) did not stimulate adequate levels of serum antibody after two 15 μg doses, and then in the USA a split vaccine produced from a clade 1 A/Vietnam/1203/2004 H5N1 virus was shown to be poorly immunogenic even after two 90 μg doses (Treanor et al 2006). However, since 2004, there has been great progress to produce and test candidate pandemic vaccines, and the influenza vaccine industry (IFPMA 2006) reports that there are at least 28 H5N1 vaccine clinical trials worldwide either currently in progress or now completed. Results from the trials are now emerging and it appears that there are some trends:

- For non-adjuvanted split vaccine, at least two doses of 90 μg are needed (Treanor et al 2006)
- For split or subunit vaccine with alum, two doses of 30–45 μg are needed (Bresson et al 2006)
- For whole virus egg grown vaccine with alum, two doses of 10–15 μg are needed (Lin et al 2006, Hehme et al 2006)
- For subunit vaccine with MF59, two doses of 7.5 μg are needed (Banzhoff 2007)
- For non-adjuvanted whole virus Vero cell grown vaccine, two doses of 7.5 μg are needed (Barrett 2007)
- For split vaccine with AS adjuvant (proprietary adjuvant of GlaxoSmithKline), two doses of 3.8 μg needed (Borkowski et al 2006).

There are of course exceptions to such trends and it was reported recently that a whole virus alum-adjuvanted H5N1 vaccine produced in Hungary is immunogenic after just one dose of 6 μg (Vajo et al 2007). Another feature of pandemic vaccine development that is causing concern is the ability of vaccines to induce immunity against variant strains of H5N1. There is evidence that clade 1 vaccines will stimulate cross-reactive antibody to clade 2 viruses albeit at reduced titres (Höschler 2006), but we do not know how this relates to cross-protection.

Apart from some uncertainties about vaccines, there are also problems associated with the techniques used to test sera from influenza vaccine trials. The haemagglutination-inhibition (HI) test is not very sensitive for H5N1 antibody unless modifications are made (Stephenson et al 2004) and both the HI test and virus neutralisation tests are not very reproducible between laboratories (Stephenson et al 2007). Furthermore, we are not sure about the correlates of immunity for pandemic influenza vaccines. There clearly is much to learn about the best use of different adjuvants and the methods employed to assess vaccine immunogenicity before the most efficient use of H5N1 vaccines can be made.

Regulatory preparedness for a pandemic

From the information available in 2004, it was apparent that a pandemic vaccine was likely to be monovalent, adjuvanted, administered in a two-dose schedule and a product of genetic modification. Such vaccines were certainly not licensed in the EU and probably were not licensed anywhere else in the world. It normally takes several months or even years to obtain regulatory approval for a new type of influenza vaccine, so plans were made in the EU to encourage each vaccine manufacturer to obtain in advance of a pandemic, regulatory approval for the type of vaccine considered necessary to combat pandemic influenza. Such approval could be obtained with a 'mock-up' pandemic vaccine strain and the licence could then be rapidly updated (in a matter of days) when the pandemic begins (CPMP 2004). This far-sighted approach has stimulated similar discussion around the world, so that pandemic vaccine licensing procedures are now being developed elsewhere. In the EU, GlaxoSmithKline was the first company to obtain a 'mock-up' pandemic licence approval in December 2006 and in April 2007, Sanofi Pasteur obtained regulatory approval for their H5N1 vaccine from the FDA in the USA.

The WHO has also stimulated regulatory discussion on various issues surrounding pandemic vaccine licensing and testing and after two workshops held in Canada and the USA in 2006; guidance will shortly be published by the WHO.

Conclusions

There has been great progress in pandemic vaccine development over the past three years and much has been learned and achieved. However we are not yet equipped to provide a fast and effective pandemic vaccine for the world community. It is vital that governments and international health authorities work together with private industry to overcome many of the issues.

References

Banzhoff A 2007 Oral presentation, Third WHO meeting on evaluation of pandemic influenza prototype vaccines in clinical trials. Geneva

Barrett N 2007 Safety and immunogenicity of a cell culture (Vero) derived whole virus H5N1 vaccine: a phase I/II dose escalation study. Oral presentation at the IX International Symposium on Respiratory Viral Infections, 3–6 March 2007, Hong Kong

Bresson JL, Perronne C, Launay O et al 2006 Safety and immunogenicity of an inactivated split-virion influenza A/Vietnam/1194/2004 (H5N1) vaccine: phase I randomised trial. Lancet 367:1657–1664

Borkowski A, Leroux-Roels I, Baras B et al 2006 Antigen sparing effect of a novel adjuvant system in a split H5N1 pandemic vaccine. International Conference on Influenza Vaccines for the World—IVW2006; 18–20 October 2006, Vienna, Austria. Abstract for oral presentation H5N1-007 (106750)

CPMP/VEG/4717/03 Committee for Proprietary Medicinal Products (CPMP) 2004 Guideline on dossier structure and content for pandemic influenza vaccine marketing authorisation application http://www.emea.europa.eu/pdfs/human/vwp/471703en.pdf

Fedson DS 2006 Vaccine development for an imminent pandemic. Why we should worry, what we should do? Hum Vaccin 2:38–42

Hehme N, Kuhn A, Mueller M et al 2006 Whole virus alum-adjuvanted pandemic vaccine: safety and immunogenicity data on a vaccine formulated with H5N1. International Conference on Influenza Vaccines for the World—IVW2006; 18–20 October 2006, Vienna, Austria. Abstract for oral presentation H5N1-001 (106378)

Höschler K 2006 Cross reactivity of antibody elicited by an inactivated split-virion A/Vietnam/1194/2004 (H5N1) influenza vaccine in healthy adults against H5N1 strains. International Conference on Influenza Vaccines for the World—IVW2006; 18–20 October 2006, Vienna, Austria. Abstract for oral presentation H5N1-001 (106378)

IFPMA 2006 R&D for avian/pandemic influenza vaccines by IFPMA Influenza Vaccine Supply International Task Force (IVSITF) members http://www.ifpma.org/Influenza/content/pdfs/Table_Avian_Pandemic_Influenza_RnD_17Oct06.pdf

Lin J, Zhang J, Dong X et al 2006 Safety and immunogenicity of an inactivated adjuvanted whole-virion influenza A (H5N1) vaccine: a phase I randomised controlled trial. Lancet 368:991–997

Nicholson KG, Colegate AE, Podda A et al 2001 Safety and antigenicity of non-adjuvanted and MF59-adjuvanted influenza A/Duck/Singapore/97 (H5N3) vaccine: a randomised trial of two potential vaccines against H5N1 influenza. Lancet 357:1937–1943

OIE 2005 Manual of diagnostic tests and vacines for terrestrial animals. Chapter 2.7.12 Avian influenza. Available at http://www.oie.int/eng/normes/mmanual/A_00037.htm

Stephenson I, Wood JM, Nicholson KG, Charlett A, Zambon MC 2004 Detection of anti-H5 responses in human sera by HI using horse erythrocytes following MF59-adjuvanted influenza A/Duck/Singapore/97 vaccine. Virus Res 103:91–95

Stephenson I, Das R-G, Wood JM, Katz J 2007 Comparison of neutralising antibody assays for detection of antibody to influenza A/H3N2 viruses: an international collaborative study. Vaccine 25:4056–4063

Treanor JJ, Campbell JD, Zangwill KM, Rowe T, Wolff M 2006 Safety and immunogenicity of an inactivated subvirion influenza A (H5N1) vaccine. New Engl J Med 354:1343–1351

Vajo Z, Kosa L, Visontay I, Jankovics M, Jankovics I 2007 Inactivated whole virus influenza A (H5N1) vaccine [letter]. Emerg Infect Dis [serial on the Internet] 2007 May. Available from http://www.cdc.gov/EID/content/13/5/06–1248.htm

World Health Assembly 2005 58.5:70–72. Available at http://www.who.int/gb/ebwha/pdf_files/WHA58/WHA58_5-en.pdf

WHO 2005 Biosafety risk assessment and guidelines for the production and quality control of human influenza pandemic vaccines. Available at http://www.who.int/biologicals/publications/ECBS%202005%20Annex%205%20Influenza.pdf

WHO 2007 Early release of influenza viruses for pandemic influenza vaccine development. Available at http://www.who.int/csr/disease/avian_influenza/guidelines/earlyrelease2007/en/index.html

DISCUSSION

Webster: You have delivered a rather discouraging message. Is there any encouragement at all?

Wood: There has been some progress. The regulatory approval process has been a huge development, whereas a few years ago we were in a mess. We are learning about the use of alum in vaccines: we know it has some moderate adjuvanting capacity with H5N1 vaccines, whereas a few years ago we thought this was the way to go, because there was no intellectual property for use of alum and there were some data from GSK that with H2N2 and H9N2 it was very beneficial. Manufacturers have now tried it with H5N1 vaccines and have shown it to be only moderately beneficial. Manufacturers are gradually testing various strategies and we are learning more. We haven't defined yet how to conserve precious antigen and deliver it well to the human immune system. This is the target.

Peiris: We talk about broad cross-protection, and the GSK vaccine, but is this just higher titres giving rise to cross reaction?

Wood: I agree, if you have a higher level of antibody to a vaccine strain, it will recognize other strains better. With better adjuvants, there is also the potential for wider protection.

Webster: Have avidity measurements been done on these responses?

Osterhaus: Not really. Assays should be carried out on the ongoing trials to really look at the breadth. For some adjuvants there are indications that they might give broader protection. I want to make a point about alum, because alum is not alum. It is not standardized. Some of the differences we see between the results of different companies may be related to the constellation of the alum. Some companies are claiming that they can do better than alum.

Wood: This may explain the Hungarian results: they have been using an alum-based whole-virus vaccine for many years in Hungary. They worked hard at it to get the adjuvanting conditions right to stimulate an effective immune response with their seasonal vaccine. They moved over to H5N1 and it worked well.

Osterhaus: This is efficacy and effectiveness. We have to be careful because there are also the safety aspects.

Wood: There are different aluminium salts. The binding of flu antigens to different alums varies according to how the antigen is prepared.

Osterhaus: What is the strategy going to be? Are we going to prevaccinate? Are we going to sit and wait?

Wood: The approach taken in the UK is to stockpile a certain amount of vaccine but not enough to immunize the whole population.

Osterhaus: There are different strategies in different countries. Stockpiling vaccine before we have the next stage is risky. If we have a number of adjuvants we could use and which give a sufficiently broad response within the subtype then we could stockpile this.

Wood: Yes, but we need the data before we can jump.

Osterhaus: Stockpiling antigen as is being practised in the USA is an interesting option, but this needs to be tested in conjunction with the adjuvant you want to use, in clinical trials.

Webster: In the longer term we need to think about H5, H7 and H2.

Osterhaus: You aren't going to stockpile all of these are you?

Wood: You identify the greatest threat and stockpile vaccine for that. For the lesser threats you just have libraries of viruses.

Osterhaus: At the moment, the problem is we don't know what the biggest threat is. I'm not sure you can ever get the knowledge, because flu is unpredictable. Stockpiling of adjuvant is a good idea if you have the technology to produce the antigen fast enough. It can be done in 4 months now.

Wood: I think 4 months is too long. If you stockpile adjuvant alone, that is not enough.

Osterhaus: You didn't mention live vaccines. Are they important?

Wood: They offer a lot of potential in the case of a pandemic. In the lead up to a pandemic, there are safety concerns about the widespread use of live vaccines. This is why there hasn't been a lot of progress. The initial results that were presented in Geneva with the Russian H5N2 vaccine and the US H5N1 vaccine were a little bit disappointing. The virus seemed to be too attenuated to replicate well, so the antibody responses were correspondingly very weak.

Peiris: With regards to dose sparing, the issue of intradermal vaccinations as opposed to intramuscular ones is interesting. You can clearly have dose sparing with seasonal vaccines when given intradermally. Is there good data for H5N1 vaccines?

Wood: Intradermal vaccination has been shown to work for seasonal flu. It is very dose sparing. Delivering the antigen to the dendritic cells seems to be an efficient way of presenting antigen. The preliminary results with H5 were disappointing, but we do not know whether this is because of the method of delivery, or because the antigen was weakly immunogenic.

Peiris: Was only one trial done?

Wood: Just one was reported.

Osterhaus: New devices are being developed. It could well be that the method of application is critical.

Anderson: In addition to the preparation of vaccines, if the pandemic occurred, getting data quickly on the vaccine used in the field would be important in terms of safety and efficacy. I assume WHO is thinking about this.

Wood: Yes, there are plans for post-marketing surveillance. This is important, especially if you have limited data before licensing the vaccine. Rob Webster, I'd like to ask about the original antigenic sin data. Were these data based purely on prior infection, or were there some based on immunization?

Webster: Prior infection only.

Wood: So we don't know whether the immune system will be channelled by immunization.

Webster: There is evidence of original antigenic sin after first infection but less information about first exposure to the antigen by vaccination.

Osterhaus: Those are limited data. I am not sure it is a real effect. The adjuvants will probably solve the problem. For unadjuvanted vaccines you may have a problem. Vaccination with the adjuvants shows that there is a very broad activity anyway. It might be a non-discussion with the proper adjuvants. We need the information.

Wood: There's also the concern that if you use an alum-based vaccine it will channel down a Th2 pathway predominantly: does this programme your immune system to induce Th2 responses to subsequent exposure?

Osterhaus: This can be looked at in the post-marketing surveillance that we discussed earlier. It is needed during the pandemic, not just after vaccination. The Guillain-Barré study was just the vaccination proper, but if you were to get a situation like measles and the RSV vaccine, with the formaldehyde-inactivated alum adjuvant, you are right in the pandemic and there is trouble.

Skehel: Is there an official CDC statement on Guillain-Barré induction in the swine flu affair? Do they accept that the vaccine delivery was directly responsible for Guillain-Barré in some people?

Anderson: I think the data are good. The clustering in time after vaccination is convincing.

Tambayh: There were also epidemiological data published after that episode. The CDC also reviewed these data (Sencer & Millar 2006).

Osterhaus: The problem is that after natural infection with flu, there can be a predisposition to Guillain-Barré as well.

Webster: The quality of the current vaccines is considerably better.

Osterhaus: With the new generation adjuvants we need to do an intense post-marketing surveillance. There is always a theoretical risk you are doing something wrong.

Wood: They are not being used on a large enough scale to detect this.

Osterhaus: Yes, Guillain-Barré could only be detected because they started to vaccinate millions of people.

Smith: What are the experimental and regulatory tests that should be done to address the question of original antigenic sin? One could do the experiment in ferrets. This wouldn't necessarily tell us what will happen in the human. I wonder about the mock dossiers that are being produced: I am pretty sure they don't include testing a second heterologous vaccination.

Wood: Some people who have received a clade 1 vaccine could be reimmunized with clade 2 and you can then see what the specificity of the antibody is.

Smith: Is there any regulatory body that has the responsibility for the use of the vaccine in this way?

Wood: No. This is a research idea that the regulators probably will not pursue.

Osterhaus: The data are thin at the moment that there is such a thing as antigenic sin. Look at the annual vaccination of the elderly: there are reasonably good results and no indication of antigenic sin.

Skehel: It is clear that if you miss out vaccinating nursing homes, there is increased mortality. It might not be the greatest vaccine, but it works.

Smith: The strains of H3 we are vaccinating with today are hugely antigenically different from 1968.

Osterhaus: These people are being vaccinated on an annual basis.

Smith: As original antigenic sin is related to the first infection, there could be a difference if a subsequent infection were antigenically close or distant to the first infection.

Skehel: The repeated vaccination every year is already living with that possibility.

Smith: Yes, but it could be a different situation when you are very close to the first infection. We don't have the experience of vaccinating every year in a large number of people close to a pandemic.

Osterhaus: This will be addressed in the animal experiments. We will get the data, and I am optimistic that with the proper adjuvants we won't see this antigenic sin.

Wood: Geoffrey Schild (Schild et al 1977) looked at sera from people who had been immunized with H3N2 vaccine. He looked at strain-specific and cross-reactive antibody to the vaccine strain and to a 1968 H3N2 virus. He found that most of the specific antibody was back to 1968, but there was a small element of specific antibody that was accumulated at the time of vaccination, which is the key thing.

Webster: Studies on original antigenic sin in animal models show that increasing the dose of antigen will obviate this problem (Fazekas de St Groth & Webster 1966, Webster 1966).

Osterhaus: The problem is that all children will be exposed, all the time. It is a difficult group to study.

Peiris: There was a study in Christchurch hospital in the 1970s which suggested that the children who were vaccinated did worse. The counter argument to that of John's would be that flu vaccine is good but not fantastic. Perhaps the reason it is not fantastic is because of the original antigenic sin?

Osterhaus: If we look at efficacy data in children under one year of age, there are no data showing that the vaccine works. If you do a monkey experiment and vaccinate a naïve monkey with the classical vaccines, there is no protection to a later challenge. In naïve children I would be surprised if the vaccine would do much.

Children over two years of age have been exposed once and you can then boost the response. If we extrapolate from animal models I am not sure the vaccines used today in naïve animals are very good. We need adjuvants in the non-primed population. On the other hand, if you use adjuvants in people already primed, such as the elderly, it is difficult to find a surplus effect.

Webster: There is also the question of the immunogenicity of H5N1 vaccine. A/Vietnam/1203/04(H5 N) 1203 is a lousy antigen in ferrets and humans, and if you take the previous variant, A/Hong Kong/213/03(H5N1), this is an excellent antigen. What is the difference between these? What is a good immunogen? We don't know.

Osterhaus: Does this really matter with the adjuvant vaccine?

Webster: Perhaps if you put the adjuvant on top of the requirements you get even more dose sparing.

Osterhaus: It could well be, but we don't know the full story.

Hoffmann: What is the science behind the mechanisms of how adjuvants increase immunogenecity?

Peiris: If these adjuvants are so good, why not use them for seasonal flu vaccines?

Osterhaus: I'm not saying we shouldn't. The benefit of developing adjuvants for pandemic vaccines is that there is a spin off to the seasonal vaccines. For many of the adjuvants that have been tested it is difficult to increase their response for those who have been primed before. There are indications, though, that some of the adjuvants that are being used for pandemic vaccines might do much better than the existing ones and could be used for seasonal flu.

References

Fazekas de St Groth S, Webster RG 1966 Disquistions on original antigenic sin. II. Proof in lower creatures. J Exp Med 124:347–361

Schild GC, Smith JWG, Cretescu L, Newman RW, Wood JM 1977 Strain specificity of antibody to haemagglutinin following inactivated A/Port Chalmers/1/73 vaccine in man: evidence for a paradoxical strain-specific antibody response. Dev Biol Stand 39:273–281

Sencer DJ, Millar JD 2006 Reflections on the 1976 Swine Flu Vaccination Program. Emerg Infect Dis 12:29–33

Webster RG 1966 Original antigenic sin in ferrets: the response to sequential infection with influenza viruses. J Immunol 97:177–183

FINAL DISCUSSION

Kahn: I have a question about the immune response to N1. Is there any evidence that there is cross-reactive antibody between the N1 in H1N1, and N1 in H5N1, and will this confer any protection? If a pandemic were to occur with H5N1, should we be giving vaccine against N1?

Webster: Richard Webby in our group has done those experiments in a mouse model. He showed that if you immunize mice with the New Caledonia neuraminidase and boost, and then challenge them with the highly lethal A/Vietnam/1203/04(H5N) 50% of the mice are protected (Sandbulte et al 2007). If you look in human serum for neuraminidase antibodies that react with Vietnam, 10–15% have antibodies.

Osterhaus: Wouldn't it be much more likely that any cross-protection you would get is related to pre-existing T cell responses? Cross-reactive cytotoxic T lymphocyte (CTL) responses are probably more efficient there.

Skehel: That is not the result of the natural experiment. If it was then you wouldn't have had the Singapore pandemic switch to the Hong Kong pandemic, since six gene products were shared. You might say that it would have been more severe if you didn't have the CTLs.

Osterhaus: Look at the H1N1 1918 flu with a W shaped curve.

Skehel: You don't know what was before that.

Osterhaus: We haven't addressed this extensively, but with the new vectors we have, we could go for an N1 or nucleoprotein or even haemagluttinin expressing vector. You could do this quite rapidly. This is an area that is largely neglected by vaccine development.

Skehel: If you could get a vaccine that was cross reactive within a sub type, this would be a real advance.

Anderson: In the context of a pandemic, if you don't have H5N1 vaccine ready, would the present H1N1 vaccine provide some protection.

Webster: One of the difficulties is that we have no standardization of neuraminidase (NA) content in seasonal influenza vaccines.

Wood: We don't know what the NA content of those vaccines is.

Webster: The other aspect that Ab Osterhaus raised was the fact that we use classical PR8 backbones, and perhaps we need to have more relevant T cell epitopes in there.

Wood: If we are targeting the activation of T cells as a vaccine strategy, is this asking for trouble? Think about the nature of the pathogenic H5N1 infection: it involves hyper activation of cytokines. If we stimulate T cells by use of vaccine, are we asking for trouble when vaccinated people are ultimately exposed to H5N1?

Osterhaus: The idea is that your T cells will knock out infected cells long before this.

Skehel: Inactive vaccines won't induce CTLs.

Peiris: You are suggesting that there are H5-specific responses that are involved? Otherwise, we are boosted most years by CTLs to flu. Why isn't that sufficient?

Skehel: It doesn't work, or there would be no more pandemics.

Osterhaus: In natural infections it is insufficient. If it is a reassortment that is a completely new avian virus, it might be that we have T cell epitopes that are unique, that have not been in the population before. Every five to 10 years we are infected with the flu if we are not vaccinated: why don't we get it every year? I think T cell immunity plays an important role there as well.

Webster: T cell immunity is important, and this was illustrated by equine flu in South Africa. Equine 2 virus had not been into Africa. In the horse population it causes some respiratory problems, but when it got into South Africa for the first time where there was no background immunity, it was a killer in horses and donkeys. Since then it has gone back to standard equine flu-causing respiratory infections.

Osterhaus: You can't discriminate between T cell and antibody-mediated responses there. This is the problem.

Peiris: What do you think of children one to two years of age who get infected with flu for the first time? They have no antibody or T cell immunity, but they don't die.

Osterhaus: It is very severe in the young children. Kids are not prone to dying. If you get it in the elderly it is a killer, but the young children have very severe disease. If they are lucky they get their first infection in the presence of maternal antibody, which mitigates the infection.

Peiris: Young children who get RSV have much more severe disease than adults.

Osterhaus: The maternal antibodies are too low to protect in those children who have it seriously. If they have an antibody titre that is high enough they only have upper respiratory infection. If they have lower maternal titres they can have pneumonia.

Anderson: There is a correlation with high neutralizing antibody titre and protection from more serious disease with infection. Not everybody with a high titre, however, is protected from serious disease.

Osterhaus: In the very young children, are they really hit in the first three or four months? At this stage the T cell responses are immature.

Anderson: You do get repeated infections, and with repeated infections there is some decrease in severity of disease. But you can get severe disease throughout life, even after multiple infections.

Osterhaus: These infections are spaced out, though.

Anderson: That's usually the case, but there are reports of repeat infections with relatively short time spans between infections.

Osterhaus: You don't see two severe influenza A infections one after the other within a couple of months.

Skehel: Is genetic engineering in terms of vaccine production going to be a major limit on what can happen in some countries?

Wood: I think that has been cleared up. When manufacturers first realized they'd have to try to get approval for using genetically modified organisms (GMOs) they threw up their hands in horror, but by now most of the vaccine manufacturers have approval to work with GMOs.

Osterhaus: The production of live vaccines is still an issue in Europe. You can use GMOs to produce vaccines, and if it is inactivated the issue is over. But if you are considering live attenuated vaccines then there is a problem in Europe.

Webster: Will this change with time?

Anderson: I would wonder if the concern about GMOs may change in the context of a pandemic!

Webster: This is the stage in the proceedings where I'd like to sum up. We have had two and a half days of presentations and active discussion on novel and re-emerging respiratory viral diseases at the beginning of the 21st century. Our focus has been on SARS and influenza, but the overarching concern throughout the meeting was the reservoirs of as yet unidentified novel viruses in lower animals and birds with the potential to spread to humans. Our continued focus has been on disease agents of humans, however, throughout the meeting the importance of the virus gene pool in lower animals and birds was emphasized. The viruses that have been most successful as zoonotic agents of disease are the RNA viruses; the properties of these viruses and their capacity for rapid evolution make them optimal agents for rapid adaptation to new hosts with the potential to cause disease. There are unidentified agents with the potential for disease in humans, and there is a plethora of zoonotic agents that could come across. These are mainly RNA viruses. We have heard about what it is that makes these optimal agents for being disease-causing agents. If we turn to SARS, this was a tragedy for those infected, but it taught us many lessons. These lessons have been listened to, particularly in Asia. I was in Asia at the time of SARS and saw the dramatic effect on economies; the chief operating officer in Hong Kong told me to give him anything to get rid of SARS. The scientific community gave him what he needed in terms of understanding that virus, but the real issue was that the virus wasn't highly transmissible: we were lucky. One of the lessons we have heard again and again is that the traditional methods of hygiene and biosafety and quarantine actually worked for SARS. These factors have been put in place in many countries, to be used if this virus reappears. However the participants agreed that the transmissibility of influenza precludes control by quarantine. Another issue that came from the SARS studies is that bats may be the primary source of many coronaviruses, and there are many more out there to revisit us in time. If we turn to influenza we come back to the topic of animal reservoirs and infections. There is a wide range of hosts, but the host we choose to act as a model depends on the questions we want to ask. For studying 1918 and pathogenesis in humans the macaque is optimal. On the other hand, the mouse has important aspects when it comes to genetics. If we turn to the structure of the haemagluttinin, it is still providing insight into function, and there are still options for drug design with the insight into that structure. If we turn to reverse genetics: this does have the potential to provide high-yielding vaccines, and this will be important for the future of vaccines for humans and animals. We have heard that the optimal vaccine strains are not fully elucidated at this time. We have also heard of some of the features that we can incorporate into vaccine strains both with point mutations of genes of neuraminidase and haemagluttinin to produce much higher-yielding strains. If we turn to vaccines and pandemics, this is one of the key issues that is being considered for the control of emerging pandemics. We have been round and round on the issues of stockpiling, on global shortages of vaccines and adjuvants, and maximizing the dose. From John Wood's presentation it is quite clear that vaccines do work. They are still the best option, but there are many questions surrounding the use of vaccines, and in my book the use of vaccines for lower animals is a key issue. The future of the control of the emerging diseases does lie in the use of vaccines.

If we turn to the key issues of whether or not an H5N1 pandemic will happen, it comes down to the molecular understanding of pathogenicity and transmissibility. We are starting to get information on pathogenicity. It is a complex issue involving multiple genes. We are not nearly so far on in our understanding of transmissibility. This involves the host as well as the virus. My own interpretation is that H5N1 is still an eradicable disease. The ultimate reservoirs have to be resolved. I have personal opinions that these are the domestic waterfowl. As we discussed earlier, there are strategies for vaccinating these populations: it is a huge task, but it could be done.

Holmes: What information would you need to identify the reservoir for H5N1?

Webster: We need prospective surveillance in the wild bird population and the domestic bird populations. We are still in a situation where we respond to outbreaks of disease in poultry. There is a lot of resistance to weekly surveillance in poultry markets, such as occurs in Hong Kong. This would tell us what the reservoirs of these viruses really are. The other surveillance is the kind of surveillance that Ab Osterhaus is doing in Europe on the wild bird populations. Are the highly pathogenic viruses being perpetuated in those wild bird populations? The evidence so far is that they are not.

Osterhaus: I fully agree on the need to have more meetings on the issue of vaccination. It is also important to bring into that equation the whole issue of antivirals. In humans, antivirals are a first line of defence: to what extent will they give us protection against the first wave of a pandemic?

Webster: Antivirals are absolutely critical and would be the first line of defence—this meeting did not focus on this aspect. In addition, there is the urgent need for future studies on combinations of these drugs, which is being largely overlooked.

Reference

Sandbulte MR, Jimenez GS, Boon AC, Smith LR, Treanor JJ, Webby RJ 2007 Cross-reactive neuraminidase antibodies afford partial protection against H5N1 in mice and are present in unexposed humans. PLos Med 4:e59

Contributor Index

Non-participating co-authors are indicated by asterisks. Entries in bold indicate papers; other entries refer to discussion contributions.

Subject Index

O

P

Q

R

S